Also by Alan Lightman

The Miraculous from the Material

The Transcendent Brain

Probable Impossibilities

Three Flames

In Praise of Wasting Time

Searching for Stars on an Island in Maine

Screening Room

The Accidental Universe

Mr g

Song of Two Worlds

Ghost

The Discoveries

A Sense of the Mysterious

Reunion

The Diagnosis

Dance for Two

Good Benito

Einstein's Dreams

A Modern Day Yankee in a Connecticut Court

Time Travel and Papa Joe's Pipe

Also by Martin Rees

If Science Is to Save Us

The End of Astronauts (with Donald Goldsmith)

From Here to Infinity

Our Final Hour: A Scientist's Warning

Our Cosmic Habitat

New Perspectives in Astrophysical Cosmology

Just Six Numbers

On the Future

Before the Beginning

Gravity's Fatal Attraction: Black Holes in the Universe (with Mitchell Begelman)

Perspectives in Astrophysical Cosmology

Cosmic Coincidences (with John Gribbin)

THE SHAPE OF WONDER

THE SHAPE OF WONDER

How Scientists Think, Work, and Live

Alan Lightman and
Martin Rees

PANTHEON BOOKS, NEW YORK

FIRST HARDCOVER EDITION
PUBLISHED BY PANTHEON BOOKS 2025

Published by Pantheon Books, a division of Penguin Random House LLC, 1745 Broadway, New York, NY 10019.

Pantheon Books and colophon are registered trademarks of Penguin Random House LLC.

Library of Congress Control Number: 2024024705
ISBN 9780593702024 (hardcover)
ISBN 9780593702031 (ebook)

penguinrandomhouse.com | pantheonbooks.com

Printed in the United States of America
1 3 5 7 9 10 8 6 4 2

The authorized representative in the EU for product safety and compliance is Penguin Random House Ireland, Morrison Chambers, 32 Nassau Street, Dublin D02 YH68, Ireland, https://eu-contact.penguin.ie.

AL dedicates this book to his wife, Jean

MR to his wife, Caroline

CONTENTS

THE SHAPE OF WONDER

Chapter I

DISCIPLINED WONDER

Many years ago, Alan took his then two-year-old daughter to the ocean for the first time. As he remembers, they had to walk quite a distance from the parking lot to the point where the ocean slid into view. Along the way, they passed various signs of the sea: sand dunes; seashells; sunbaked crab claws; delicate piping plovers, which would run and peck, run and peck, run and peck; clumps of sea lavender growing between rocks; sea glass; and an occasional empty soda can. The air smelled salty and fresh. His daughter followed a zigzagging path, squatting here and there to examine an interesting rock or shell. Then they climbed over the crest of a final sand dune. And suddenly, the ocean appeared before them, silent and huge, a turquoise skin spreading out and out until it joined with the sky. He was anxious about his daughter's reaction to her first sight of infinity. Would she be frightened, elated, indifferent? For a moment, she froze. Then she broke out in a smile.

Child psychologists have long noted that curiosity and wonder bubble up naturally in young children, even if these emotions often dwindle with age. Indeed, the drive to explore and understand the world around us must have had survival benefit in our evolutionary history. The capacity for wonder is part of our DNA.

As a matter of fact, we may think of the scientific enterprise as disciplined wonder. Good scientists view the physical world with the curiosity and wonder of children, but with additional training and tools to understand stars and rainbows and spiderwebs in precise terms.

One recent afternoon, we were visited by a high-school English teacher named Lucile. For many years, Lucile taught American and British literature and won a statewide award for excellence in teaching. She's a poet herself. As we walked along a winding path through the woods, we talked about the recent wildfires in Canada, the COVID pandemic and the vaccines to fight it, and the vast new capabilities of artificial intelligence. Lucile had much to say. Our conversation turned to the remarkably fast pace of science and technology. Lucile was extremely interested in all the new developments but also a bit uneasy about the future. After a while, she made the point that she really doesn't know what scientists do. How do they spend their time, in the lab and at home? Do they consider themselves different from other people? Finally, Lucile commented that scientists always seem to be changing their minds—on the foods we should eat, on the medicines we should take, on how often women should get mammograms, even on the number of planets in the solar system.

This book addresses Lucile's concerns. It strikes us that the lack of understanding of the daily activities, thinking, and motivations of scientists may be a factor in the mistrust of scientific information in some of the public today—and in the feeling that many scientists are out of touch with their societies. The mistrust is especially evident when scientists are seen to be the mouthpieces of governments and state institutions, motivated by financial or political interests. In that case, scientists are considered part of the "elite establishment," disconnected from the lives of ordinary people. Such mistrust is part of the worldwide populist movement. We recently saw this mistrust during the COVID pandemic, in the resistance to government-mandated masking and vaccines and the skepticism toward the recommendations of national health organizations. Not long ago, we also witnessed such mistrust in the refusal to acknowledge human-caused climate change. The problem may not lie with individual scientists themselves. A recent report of the Edelman Trust Barometer, an international survey by the Edelman global communications company, suggests that people seem to trust scientists as individuals but not as representatives of governments and other state institutions. That view was confirmed by a recent study published in *Nature* that involved 68 countries. Most people in most countries trust scientists themselves. However, it is often difficult to separate scientists from their institutions, leading to skepticism about their motivations and recommendations.

Speaking further to Lucile's concerns, much of the public does not understand the nature of the scientific enterprise, especially the role of revision. When scientists change their recommendations, it is not because they are wishy-washy. It is because new information has become available. You wouldn't keep recommending a particular model car to your friends after you found out it had brake defects.

The mistrust of scientists and their institutions, for whatever reason, is an urgent problem. Science and technology play a crucial role in modern life. Climate change, the manipulation of DNA to cure disease or create altered human beings, computers that can replace jobs or make decisions on the battlefield as well as in daily life are all careening toward us with exponential speed. Science is changing the world. We must understand the science, and we must trust its practitioners. We ignore them at our peril.

Undoubtedly, our leaders and lawmakers will need to address the challenges—both the benefits and the dangers—of these rapid new developments. And ultimately, our elected officials respond to citizens and voters. In turn, it is the responsibility of those citizens to have some understanding of science and scientists, how they think and work.

In this and the following chapters, we have tried to humanize scientists and their world. We want to show scientists as individuals, concerned about their societies as all of us are. We will portray the professional and personal lives of a range of scientists, what they do, how they think, their responsibilities to their societies. And we'll explore the nature of the scientific enterprise, especially the role of critical thinking and the manner in which scientific theories and recommendations are revised with the arrival of new information and evidence. Addressing the mistrust of the *institutions* of science and the larger mistrust of state institutions is beyond the scope of this book, and the needed social and institutional changes may take a long time. What we can do here is to give an honest picture of scientists as people and how they work and think.

We have tried as much as possible to listen to scientists speaking in their own voices, so there are many direct quotations and profiles. Above all, we want to demonstrate and affirm that scientists are ordinary people, albeit with special training and skills. Although scientists are intelligent, determined, and almost always passionate about their work, they are also interested in other things: they have hobbies and families; they enjoy cooking and eating; they go to the cinema and sports events. And their professional way of examining the physical world, often called the "scientific method," is really just critical thinking, which can be found in the work of attorneys, doctors, accountants, auto mechanics, and others.

Many of us envision the scientist as the older Einstein, a mystifying genius with wild frizzy hair, his head in the clouds, a slide rule (or pocket calculator) clipped to his baggy sweatshirt, a creator of arcane theories. That stereotype fits few scientists. Scientists don't fall into a single mold. First of all, scientists are women as well as men. Some sit and think; some build instruments and do experiments; some work alone, others work in small groups, and some in enormous consortiums; some do research and teach at universities, others work for industry or for big companies; some work on pure problems with no foreseeable social benefit, others on problems with direct application to improving the lives of people; some scientists are married with children, others not; some come from Europe or Asia, others not. Each scientist chooses the particular kind of science and working environment that suits her individual temperament as much as her skill set. All are human beings, with passions, jealousies, wrong turns, hopes, and anxieties like everyone else.

Furthermore, there is no single scientific personality type. Scientists can be bold and self-confident revolutionaries, like the New Zealand physicist Ernest Rutherford, who discovered the nucleus of the atom, or James Watson, one of the co-discoverers of the structure of DNA, or the marine biologist

Rachel Carson, who helped launch the environmental movement. They can also be modest and diffident, like Charles Darwin, or the Austrian physicist Lise Meitner, who made significant contributions to the understanding of nuclear fission, or Alexander Fleming, who discovered penicillin. The British physiologist William Bayliss, who discovered the first hormone in 1902, was cautious, meticulous, and in love with the details, while his collaborator, Ernest Starling, was brisk, impatient, and engaged mainly with the broad sweep of things. What all of these men and women shared was a sense of wonder at the natural world, a curiosity to know, a sheer pleasure in solving puzzles, and an independence of mind.

Let's return to science as disciplined wonder. Scientists rarely discuss such personal feelings as curiosity and wonder in professional academic papers, but we can find these more intimate confessions in their letters and popular books. In his autobiography, the physicist Werner Heisenberg (1901–1976) described the transcendent moment when he realized that his new theory of quantum mechanics would succeed in explaining the hidden world of the atom. At the end of May 1925, after months of struggling with his theory, he fell ill with hay fever and took a two-week leave of absence from the University of Göttingen.

> I made straight for Heligoland, where I hoped to recover quickly in the bracing sea air. . . . Apart from daily walks and long swims, there was nothing in Heligoland to distract me from my problem. . . . When the first terms [in the mathematical equations] seemed to accord with the energy principle, I became rather excited, and I began

> to make countless arithmetical errors. As a result, it was almost three o'clock in the morning before the final result of my computations lay before me. . . . At first, I was deeply alarmed. I had the feeling that, through the surface of atomic phenomena, I was looking at a strangely beautiful interior, and felt almost giddy at the thought that I now had to probe this wealth of mathematical structures nature had so generously spread out before me. I was far too excited to sleep.

Another example of scientific wonder: In 1896, it was found that uranium produces rays of great penetrating power, now known as radioactivity. However, the nature and origin of these rays was mysterious. Soon after, Marie Curie (1867–1934)—working with her husband in a poorly vented shed next to the City of Paris Industrial Physics and Chemistry Higher Educational Institution—did experiments to show that the puzzling rays were coming from within the atoms, evidently spitting out tiny pieces of themselves. Here was the first indication that the atom, for two thousand years the embodiment of indivisible unity, was not indivisible. Curie went on to discover two new chemical elements, polonium (named after her native Poland) and radium. Curie describes her career in science in this way: "I belong in the ranks of those who have cultivated the beauty that is the distinctive feature of scientific research. A scientist in the laboratory is not just a technician; he confronts the laws of nature as a child confronts the world of fairy tales. . . . If there is something vital in everything that I notice, this is the spirit of adventure which seems inextinguishable and is bound up with curiosity."

What do we mean by discipline? Of course, there is the

usual understanding of discipline: a particular area of investigation, which, in science, translates to the useful segregation of knowledge into the different fields of biology, chemistry, physics, astronomy, and so on. These segregations help focus the particular tools of inquiry. Another part of discipline is the purposeful building of devices to study nature, ranging from the microscope to the telescope to the world's largest scientific instrument, the giant subatomic particle accelerator named CERN near Geneva, Switzerland, an underground ring stretching 17 miles in circumference and employing thousands of scientists worldwide. Another part of discipline is the pointed and methodical investigation of a problem.

Building his own lenses and microscopes, the Dutch scientist Antonie van Leeuwenhoek (1632–1723) was one of the first people to discover and observe the microworld, invisible to the human eye. We can sense his excitement and joy, as well as the discipline of his observations, in a letter Leeuwenhoek wrote to the Royal Society of England on September 7, 1674: "Among these streaks there were besides very many little animalcules. . . . And the motion of most of these animalcules in the water was so swift, and so various upwards, downwards and round about that 'twas wonderful to see: and I judged that some of these little creatures were about a thousand times smaller than the smallest ones I have ever yet seen upon the rind of cheese."

Discipline also includes the methodical search for patterns and regularities in natural phenomena, often resulting in quantitative laws. One of the first human beings to formulate a law of the physical world was the Greek mathematician and scientist Archimedes (ca 287 BC–ca 212 BC). Around 250 BC, Archimedes formulated his law of floating bodies: "Any solid lighter [less dense] than a fluid will, if placed in the fluid, be so far immersed that the weight of the solid will be equal to the weight of the fluid displaced."

We can speculate on how Archimedes arrived at his law. At the time, balance scales were available for weighing goods in the market. The scientist could have first weighed an object, then placed it in a rectangular container of water and measured the rise in height of the water. The area of the base of the container multiplied by the height of the rise would give the volume of water displaced. Finally, that volume of water could be placed in another container and weighed. Undoubtedly, Archimedes performed this exercise many times with different objects before devising the law. He probably also performed the experiment with other liquids, like mercury, to discover the generality of the law. In other words, he had discipline.

Here's a wonderful law of nature that you can verify for yourself: Drop a weight to the floor from a height of 4 feet and time the duration of its fall. You should get about 0.5 seconds. From a height of 8 feet, you should get about 0.7 seconds. From a height of 16 feet, about 1 second. Repeat from several more heights, both bigger and smaller, and you will discover the result that the time exactly doubles with every quadrupling of the height, a rule first found by the Italian physicist Galileo (1564–1642). If you draw a graph of the numbers, time against distance, you will find that they form a boomerang-like curve called a parabola, familiar to high-school students. With the picture, you can now predict the time to fall from any height. You have discovered firsthand a fundamental law about forces and falling bodies.

—

Today, there are some 8.8 million people worldwide called scientists. Scientists differ from engineers. Scientists explore the natural world and are mainly interested in understanding and describing it. Scientists have many motivations, which we will discuss in later chapters, but a principal one is simply to learn the how and the why of the physical world around us. Scientists study such things as the way that various chemicals combine to produce new substances, the manner in which cells in living organisms receive and produce energy, the insides of atoms, the structure of galaxies.

By contrast, engineers usually begin with a human problem and build things to solve that problem: for example, houses and bridges, automobiles for fast transportation, water and waste systems, and computer processors. Engineering is applied science. Engineers transform science into material things designed to improve the quality of life. We should point out, however, that the boundary between applied science and engineering is not sharp. We should also point out that pure science often turns into applied science. The active ingredient of the leading vaccines used to combat COVID-19, called mRNA, was not invented or discovered in order to fight disease. It was discovered in 1961 by a team of biologists at the Institut Pasteur in Paris, trying to understand how the messages carried by DNA are "read" and then used to build proteins. The scientists identified mRNA, standing for "messenger RNA," as the responsible agent. It was years later that mRNA was found to be useful as a vaccine.

Many of the conveniences, pleasures, and necessities of modern life are the result of the marriage of pure and applied science: automobiles, computers, cell phones, radiation treatment for cancer, antibiotics, vacuum cleaners. Science and its applications are everywhere around us, so common as to be almost invisible, like the air.

Why do we do science? The urge to make sense of the world, lying at the foundations of science and the motivations of its practitioners, probably dates back to our ancestral beginnings. Anthropological studies of the bones of early humans (of the genus *Homo*) show that human brains increased rapidly in size between about 800,000 and 200,000 years ago. That period of time coincides with an era of large climate fluctuations. Evolutionary biologists propose that the significant increase in capacity of our brains was driven by a need to adapt to a changing environment, with clear survival benefit. Once a brain possesses such a large capacity, there will inevitably be by-products of that capacity. Stephen Jay Gould and others have called such evolutionary by-products "spandrels": traits that do not have any direct survival benefit in themselves but are by-products of traits that do. The ability to write poetry, for example, is a spandrel. The sensitivity to rhythm and sound, the basis of poetry, would have had direct survival benefit. Similarly, our cities, our machines, and our computers should be viewed as by-products of high intelligence. Most importantly, our desire to understand the cosmos and our place in it must be an outcome of our big brains, whether of direct survival benefit or not.

Making sense of the world is related to the human urge to seek patterns and order, mentioned earlier. According to the gestalt school of psychology, we human beings unavoidably tend to organize all experience into meaningful patterns. We see scorpions and lions and archers in the arrangements of stars in the sky. When a picture of random dots is presented to us, we parse it into figures and background. When we see a broken circle, we mentally complete it. According to Irvin Yalom, emeritus professor of psychiatry at Stanford University, when incoming stimuli don't form patterns, "one feels tense, annoyed, and dissatisfied."

The human need to find order in the world is reflected in the systems of fundamental elements articulated in all cultural traditions. Thinkers in ancient India conceived of a system of three elements for constructing the cosmos: fire, water, and earth. Fire was associated with bone and speech, water with blood and urine, earth with flesh and mind. For the ancient Chinese, the fundamental elements were wood, fire, metal, water, and earth. Aristotle also built the cosmos out of five elements: earth, air, water, fire, and aether (for the heavenly bodies). Today's fundamental elements are subatomic particles: quarks, leptons, and bosons.

The urge to seek order is found not only in scientists. We see it in librarians, who sort books into different categories and create catalogs; in sports fans, who gather and organize the statistics of players; in writers, who outline a prospective book before starting the first chapter. We human beings are pattern makers and we are organizers.

Besides the desire to find order and pattern, one of the motivations to do science is to improve the quality of life, to make life easier, to make communication faster, to find cures for mental illness, and so on. We will explore these and other motivations of scientists in more detail in the following chapters and in the short profiles of scientists between chapters.

—

The methodology of scientists, in which hypotheses are tested against observation and experiment, is often called the "scientific method," but it might better be called "critical thinking." Scientists certainly don't have any ownership of such thinking. As mentioned earlier, many nonscientists exhibit critical thinking. In general, we might define the scientific method as a process of asking questions, proposing hypotheses, methodically gathering information and evidence to test those hypotheses, and rejecting those that disagree with the evidence.

We recently talked about critical thinking with our friend Charlie, who owns an automobile repair shop. Charlie says that people often come into his shop with the check-engine warning on their dashboards. There's some problem with the car, but what is it? According to Charlie, the most frequent causes are a malfunctioning catalytic convertor (which transforms an engine's toxic emissions into safe gases) or a defective oxygen sensor (which figures out the optimum air-to-fuel mixture for the engine). But not always. The problem could be anything from a loose gas cap to a faulty spark plug to corroded spark plug wires. The mechanic might first check the gas cap, the easiest thing to fix. Then he or she might ask the owner if their car has been burning an unusual amount of oil, evidence that oil is leaking onto the spark plugs and corroding them. If that's not it, the mechanic moves on to the next potential problem, systematically examining and replacing, in turn, wires, spark plugs, oxygen sensors, and finally the catalytic convertor (most expensive). After each possible fix, the mechanic, of course, checks to see if the check-engine light has gone off. At some point, the mechanic may decide that the check-engine light itself is defective and replace it.

Other familiar examples of critical thinking in daily life are making family budgets and carefully monitoring expenses to make sure total expenses don't exceed income, using a thermometer to determine if a child has a fever in an attempt to diagnose an illness, using weather predictions to plan picnics, learning by trial and error which roads have the most traffic at different times of day. All of us use critical thinking at various times.

The first person often credited with putting critical thinking into practice in science in a regular way was the Egyptian physicist Alhazen (Ibn al-Haytham) (965–1040 AD). Alhazen was a pioneer in the study of light, optics, and vision. At the time, many people believed that the light required in vision originated in the eye. According to this view, the eye is like a flashlight. It sends out a beam of light and illuminates objects. With his own experiments and mathematics, Alhazen worked out a theory of the trajectory of light rays, establishing a correspondence between particular points on an external object and particular points inside the "crystalline humor" of the eye. Besides his own experiments, he sensibly observed that it would be absurd to think that light emerging from a person's eyes could illuminate the entire visible sky as soon as he opened them.

Before Alhazen, many thinkers thought that statements about the physical world could be accepted simply on their appeal or on the authority of the speaker, without the need to actually test them against nature. Since the time of Alhazen, test and experiment have been king. Sometimes reluctantly, scientists have discarded attractive theories and hypotheses because they disagreed with the findings of subsequent experiments. Examples of such once appealing but ultimately rejected theories are the miasma theory that all diseases are caused by bad air arising from the ground, the idea that heat is a fluid, the theory that miniature human beings reside whole inside eggs and sperm, and the belief that a very tenuous and transparent substance called the aether fills up space and is responsible for the propagation of light. What changed? More observation, experiment, and evidence.

The rejection or revision of theories upon being confronted with conflicting evidence is a central part of the scientific method, and all critical thinking. However, that process of unflinching evaluation is often carried out by the *community* of scientists, rather than by individual scientists. Full of the passion and emotional commitment that allows them to study a single problem for years, individual scientists sometimes become so enamored of their theories and experimental results that they lose objectivity and become blind to contradictory evidence. Rarely so in the community of scientists. Objectivity is achieved by the diverse constellation of scientists, critiquing one another's work, reading one another's papers, repeating one another's experiments. Competition and skepticism are part of the modus operandi of scientists.

Many scholars trace the beginnings of modern science, the so-called Scientific Revolution, back to the work of Copernicus (1473–1543), especially his proposal that the Sun, not the Earth, is the center of our solar system. To arrive at that theory, Copernicus used his own observations of the planets, in addition to logical contradictions he found in Aristotle's and Ptolemy's Earth-centered solar system. Following Copernicus were two leaders of the Scientific Revolution, the Italian physicist Galileo, mentioned earlier, and the great Isaac Newton (1643–1727), who formulated general laws of force and motion, worked out a mathematical expression for the force of gravity, and invented calculus. Scholars still disagree on why modern science originated in Europe and not in other parts of the world, but many non-Western thinkers, such as Alhazen in Egypt, helped build the foundations on which science rests.

Since the Scientific Revolution, the view of science and scientists has evolved and also differed from one country to the next. In Europe, science was long considered a part of culture, and a few privileged individuals could devote themselves to what was then called "natural philosophy." Isaac Newton, as a fellow of Cambridge University, needed no justification for his studies of physics. Carl Friedrich Gauss, who made brilliant contributions to mathematics and astronomy in the early nineteenth century, was supported through the patronage of the Duke of Brunswick. By contrast, when science got underway in America, in the middle 1800s, the democratic ideals of that young country demanded a direct accounting to the people, a direct benefit to society. Scientific research was usually supported only if it was part of a practical or technological enterprise, like the National Weather Service, founded in 1870, the U.S. Geological Survey, founded in 1879, or the National Bureau of Standards, founded in 1901. Even the American Philosophical Society, founded by Benjamin Franklin in 1743, had as its mission "the promotion of useful knowledge." Gradually, the United States began to take pride in and identify with its applied science and technological achievements (as opposed to pure science). Until recently, the American hero of science was Thomas Edison, not J. Willard Gibbs, who made fundamental contributions to the theory of heat; and not Richard Feynman, who did fundamental work in the quantum theory of electricity and charged particles. These days, with the computer revolution, American heroes of technology are people such as Bill Gates and Steve Jobs and Sergey Brin.

As we have mentioned earlier, although science has undeniably improved the quality of life, from antibiotics to digital computers to GPS (which would not work without taking into account Einstein's relativity), a significant portion of society today views scientists and the institutions of science with skepticism and mistrust. Science has its strengths, and it also has its limitations. On one extreme is the belief that science has all the answers, not simply to landing men and women on the Moon but how to structure governments and economies, how to decide if a murderer should receive capital punishment, and many other social, moral, and even aesthetic issues. On the other extreme, as we have said, are people suspicious not necessarily of science itself but of the institutions of science and its priests. This group associates the universities and laboratories and professors of science with the elite establishment, which is out of touch with the lives of ordinary working people and fails to understand lived human experience.

Neither of these two extremes is healthy for our world. We need some humility. Science and its practitioners can indeed answer many questions, but not all questions, and not those that concern social, moral, and aesthetic issues. At the same time, most scientists accept the value and validity of human experience, even if their institutions appear to be cold slabs of mortar and brick.

While scientists themselves are members of their societies, science itself is strictly apolitical. Science has no national allegiances, no goals other than to improve our understanding of the world and the quality of human life. Science does not have an ideology. It is a way of looking at the world, a way of understanding the world by asking questions, making hypotheses, testing those hypotheses, making predictions. Shortly after the end of World War I, which, like all wars, brought out the worst in us human beings, all nations rejoiced in a successful test of the new theory of gravity proposed by Albert Einstein. That theory was confirmed by the bending of light rays by the gravity of the Sun, as predicted, a tribute to the intellect of humankind and the nobility of our search for knowledge, a cause for global celebration. Science expressed the best in us.

Just as science itself is apolitical, it is important to emphasize that neither science nor technology have moral or ethical values in themselves. It is we humans who impose those values, in how we choose to use science and technology.

Scientists have a responsibility to their societies, both as authority figures in certain areas and as regular citizens. In all issues involving science, scientists have special expertise and a special responsibility to provide support to policymakers and help make sure that science is used for good and not for ill. On nonscientific matters, such as immigration and abortion, scientists do not have any particular expertise but have the same responsibilities as other citizens of democratic societies.

We live in a special moment. Our world today faces unique challenges, including climate change, biotechnology, new medical procedures, and artificial intelligence. Science can help confront these challenges. Beyond these physical issues are the social, ethical, and political issues. Much of the world today is deeply divided. Some of that polarization includes a widespread suspicion of traditional power centers. Scientists and the institutions of science are considered part of those power centers, aloof from the rest of us. We must try to understand scientists as people and what they actually do, how they think, work, and live.

That is the aim of this book.

Chapter II

WHY SCIENCE?

On July 8, 1943, Thomas W. Wallace, only seven months into his job as lieutenant governor of New York, came down with pneumonia. He was quickly driven to Ellis Hospital in Schenectady, where he was placed in an oxygen tent. For a while, his condition improved. But then he suffered a relapse. Despite being given an injection of serum rushed from state laboratories in Albany, a bronchoscopy procedure to clear the congestion in his lungs, and three blood transfusions, Wallace passed away a few days later, with his wife and two young children at his bedside. He was forty-three years old.

By today's standards, Mr. Wallace died young. But, in fact, people born in 1900 like Mr. Wallace had a life expectancy of only forty-seven years. Little more than a century later, life expectancy had dramatically increased to seventy-nine years—all due to a better understanding of microscopic germs as the cause of infectious diseases, public health measures, and the discovery and application of antibiotics. Indeed, if Mr. Wallace had been born only a few years later, his pneumonia might have been easily cured with the newly available penicillin, discovered in 1928 by a Scottish biologist named Alexander Fleming.

People today are not only living longer. We're wealthier. At the beginning of the twentieth century, 65 percent of the world's population lived in extreme poverty, defined by the World Bank as less than $2.47 per person per day in 2022 dollars. Today, the fraction of people worldwide living in that condition is about 9 percent.

What happened to bring about such astonishing increases in human life and well-being? Science and technology. In addition to advances in biology and medicine, brought about by such people as Louis Pasteur and Alexander Fleming, the first and second Industrial Revolutions, beginning in the mid-eighteenth century, vastly increased food production, transportation, and the creation of goods. Crop rotation, selective breeding, mechanized plows, and synthetic fertilizers increased wheat production from 0.5 tons per hectare (2.5 acres) to 8 tons per hectare today. Cyrus McCormick's threshing machine led to a 500 percent increase in wheat harvesting per hour. Isaac Singer's sewing machine lowered the time to make a shirt from fourteen hours to one hour. The newly invented electric dynamo generated power for factories and industry. The British inventor Henry Bessemer discovered a process for greatly improving the production of steel, used in plows and almost everything else. Railroads. The cotton gin. The seed drill. Tractors. Electricity.

Ideas and techniques travel fast from the lab bench to the engineering shop to the industrial warehouse. In 1832, the British physicist Michael Faraday discovered the operating principle for electromagnetic generators and motors. In 1858, the Belgian-French engineer Étienne Lenoir invented the first commercial internal combustion engine. In the period 1856–1863, the biologist and botanist Gregor Mendel discovered the laws of biological inheritance. Then, in the late 1860s, the chemist Friedrich Miescher identified the molecule DNA, eventually leading to the unraveling of its structure by Rosalind Franklin, James Watson, and Francis Crick in 1953. Today, DNA therapy is used to treat such diseases as Leber congenital amaurosis, which causes blindness, and various blood cancers such as acute lymphoblastic leukemia and large B-cell lymphoma. In the future, gene therapies may be used to treat cystic fibrosis, hemophilia, sickle cell disease, and many other diseases. In the early twentieth century, Einstein formulated equations for the peculiarly relative flow of time, now an indispensable part of the workings of all GPS systems.

We have labeled ourselves *Homo sapiens.* It means "wise humans." Over thousands of years of our history, no other activity is more worthy of that name than our science and technology. We are thinkers, we are dreamers, we are tool makers, we are inventors, we are builders, we are creators. Prometheus was chained to a rock for stealing the fire of the gods and giving it to human beings. But, in fact, we have made fire ourselves. Without help from the gods, we have created steam engines and internal combustion machines and antibiotics. If an intelligent being were monitoring our planet through a giant telescope from the far reaches of space—even without any inkling of Alexander Fleming or Michael Faraday or Albert Einstein—he/she/it would observe a remarkable transformation of the planet: first the gradual organization of the land into the neat rows of agricultural fields, then buildings, then the congregation of buildings into cities, then a proliferation of points of light on the surface of the Earth, then radio signals, then large machines flying in the air . . .

We have made our world, and we have made it with science

and technology. We're so accustomed to the science and technology around us that we hardly notice. It has become almost invisible. But consider our electric lights, our automobiles, our refrigerators and washing machines, our televisions, our computers and smartphones. As little as a century ago, none of these items were available, some not even dreamed of. Or consider the extremely rapid development of vaccines during the recent COVID pandemic. We live in a built world, and we are the builders.

The foundation for most of our medicine and technology is science. Pasteur and Fleming and Faraday and Franklin and Einstein were scientists. But beyond its many practical applications, science has changed our understanding of the cosmos and our place in it. Science has given us new ways to think about the world and about ourselves. Once we humans believed that our planet was the center of the universe. Discoveries in astronomy have shown that, instead, it is our Sun that is the center of our solar system, which itself resides on the outskirts of an ordinary galaxy, which itself is one among zillions of galaxies. Once we thought that we were different from all other living creatures, and that the world was made for our benefit. Following the ideas of Darwin, discoveries in biology and paleontology have shown that we evolved from earlier life-forms.

We *Homo sapiens,* like all other animals, are part of a grand evolutionary process that can be traced back to the beginnings of life on Earth. We are part of the overarching unity of the entire web of life on our planet. The understanding of DNA has allowed us to gaze into our very construction, removing much of the mystery of living organisms. Discoveries in brain science have shown us that our emotions, our angers and sympathies, even our personalities are controlled by chemicals within the three pounds of mushy mass in our skulls. The discovery of the expansion of the universe and its Big Bang beginnings has eliminated the long-held belief of an unchanging and eternal cosmos. The discovery of relativity and quantum physics has demonstrated that nature at high speeds and small distances behaves in ways very foreign to our sensory perceptions. Yet we have harnessed those discoveries in building our GPS systems and our computers. In effect, science and its applications have greatly extended our human sensory apparatus. Science and its applications have made us super beings.

What about our place in the cosmos, and our future as human beings? Astronomy, in particular, has provided an awareness of the huge spans of space and time in the universe and also the vast amount of time that lies in the future. Even people who accept Darwin's ideas tend to think that human beings are in some sense the culmination, the end point, of evolution. Astronomers know, however, that the time lying ahead is at least as long as the time that has elapsed up until now. In fact, according to our best thinking, it could be infinite.

We should not think of human beings as the final form of life. There is as much potential for evolution in the future as there has been to get from a bacterium to us. But it is not evolution in the Darwinian sense, driven by the forces of natural selection. We have bypassed those forces. We are modifying our evolution by our own hand. We are remaking ourselves. Right under our noses, *Homo sapiens* is transitioning into a new type of species that might be called *Homo techno.* And the change is happening not over millions of years. It's happening in single human lifetimes, by our own inventions and technology.

We have already seen small examples of such techno evolution in our eyeglasses and hearing aids. In 2015, a man named

Eric Sorto, paralyzed from the neck down, had a computer chip implanted in his brain that allowed him to move a robot arm by pure thought. And only recently, new large language models in artificial intelligence, such as ChatGPT, promise to greatly expand our ability to manipulate and employ enormous quantities of data. At the University of Cincinnati in the United States, researchers at the eXtended Reality Lab have created an embodied/virtual reality system in which people exercising on a treadmill alone in their house have the experience of being in a gym, surrounded by other people also exercising. Of course, such high-tech innovations, providing the illusion of a social environment, have both pluses and minuses.

But far more awaits in the future. In a century or less, we may have special lenses implanted in our eyes that do what our external detectors do now, allowing us to see X-rays and other frequencies of light much higher than the visible part of the electromagnetic spectrum. With this technology and off-the-shelf X-ray emitters, we will be able to see through walls and many other surfaces that ordinary light cannot penetrate. A hundred years from now, we may have computer chips implanted in our brains that connect them directly to the internet. In such a situation, we may need only think of a piece of information we need, and it will be instantly transmitted to our brains. Using the same technology, we might be able to communicate directly with the minds of other people through the internet. (Such a scenario would raise vast new issues of privacy and intellectual property and probably require new kinds of laws and legal constructions.) Fifty years from now or less, we may have tiny robots, the size of red blood cells or smaller, that can be injected into our bodies to kill cancer cells, deliver drugs with highly targeted precision, repair damaged or faulty

DNA, and greatly enhance our immune system. With developments in neuroscience and the understanding of memory storage, we may have brain implants that teach us new languages in minutes. Words in Chinese or Swahili could be streamed to our brains in real time. And if we wish to utter those words, instructions for the unfamiliar movements of mouth muscles could also be streamed in real time. And we can hardly foresee all the changes that advanced artificial intelligence will bring about, both the benefits and the dangers.

In sum, human beings in the future, *Homo techno,* will be part human and part machine. These cyborg beings will have vastly greater physical and mental abilities than we do now. We cannot imagine what we will be even a hundred years in the future, just as people in the year 1900, the year that Thomas Wallace was born, could not have imagined cell phones and antibiotics. All of these developments, now and in the future, will continue to change how we understand ourselves and our place in the world. These advances are not only material. They are also about ideas. They are part of our human culture, and it is a universal culture, transcending language and ethnicity and national boundaries.

A principal driver of science is simply the human desire to fathom this strange cosmos we find ourselves in. What are those glittering points of light in the night sky overhead? Does space go on forever? Why do we die, and what happens to us afterward? How did the universe come into being? What makes the seasons? How big is the Earth? What causes lightning? What kind of things are the Sun and the Moon?

Human beings must have been asking these questions for a

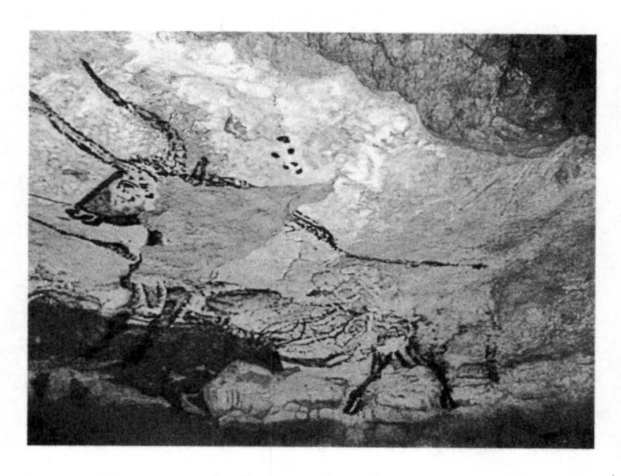

very long time. Stars and the night sky are portrayed in the earliest cave paintings. Above is one example, from about twenty thousand years ago, found in the cave system in Lascaux in southern France. According to archaeological and astronomical experts, this painting depicts the Pleiades constellation, the six dots. Other cave paintings suggest a fairly sophisticated knowledge of astronomy at that time, far earlier than Stonehenge.

Researchers at the Universities of Edinburgh and Kent figured out that many of the cave paintings mark the dates of significant comet sightings and were correlated with star constellations visible at those times. A particularly notable example, also found in Lascaux, is called the Shaft of the Dead Man, shown below.

Contemporary scientists believe this image symbolizes a comet strike that occurred around 15,200 BC. It is certainly a strange picture. The dying man, lying down, wears a bird mask. His erect phallus points toward the bull or bison on the right. A staff below him bears the image of a bird. The bison or bull signifies the current-day constellation Taurus, the researchers think, while the cave's horse (part of which can be seen on the left) represents the stars making up the Leo constel-

lation. It is reasonable to assume that the naming and depiction of constellations were our ancestors' attempts to find patterns and meaning in the resplendent night sky. Comets were surely a wondrous and baffling phenomenon.

Other attempts to make sense of the world can be found in origin stories. Every culture has had one. The oldest recorded story of creation is the Sumerian *Enûma Elish* (1900–1600 BC). Its thousand or so lines, written in Akkadian cuneiform, were found on seven clay tablets in the Library of Ashurbanipal in Nineveh, the capital of the ancient Assyrian Empire, now Mosul, Iraq. The name *Enûma Elish* comes from the first two words of the poem and means "When on high." The poem tells how in the beginning, before sky or ground, there was only Apsu, the sweet waters, and Ti'amat, the salt waters. "When above the heaven had not yet been named / And below the earth had not yet been named . . ." In time, these stretched into the giant ring of the horizon. The gods seethed in disorganized chaos inside the body of Ti'amat. At some point, the hero of the story is born, Marduk, a god of storm and thunder. Marduk does battle with Ti'amat and her army. Eventually, Marduk triumphs and gives birth to the gods. We might conjecture that if these people had a concept of "laws of nature," such laws would have originated with Marduk's action, or perhaps they were embodied in Apsu, the prime mover.

A leading Chinese origin story is the myth of Pangu, at least as old as the time of the Three Kingdoms (220–280 AD). In the beginning, for unmeasured time, the universe was formless, just as in the *Enûma Elish.* Then it coalesced into a "cosmic egg" for about eighteen thousand years. Within that egg, the opposite principles of yin and yang (the foundation of Chinese thought) accommodated each other. Like Marduk in the *Enûma Elish,* Pangu created the world by separating yin and yang with his axe. Yin became the earth and yang the sky.

In his *Timaeus* (360 BC), Plato argued that a Creator made order from disorder, as in the Marduk and Pangu stories. The operational standard was "the good," some perfect, eternal, and changeless ideal, existing outside time and space.

The human desire to make order out of the jumble of forces and phenomena of the world, evident in all of these early cosmologies and many more, must have satisfied a deep psychological need. That need must also be part of the impulse to do science. One can see the impulse in the orderly arrangement of massive stones at Stonehenge, or in the Vikram Samvat calendar of ancient India, or in Isaac Newton's *Principia.*

We might compare the ancient Sumerian and Chinese stories of the origin of the cosmos with the story told by contemporary physicists, called the Big Bang: Our universe began some 14 billion years ago in a state of extremely high density and temperature. We can trace cosmic history with confidence back to one-millionth of a second after the beginning, when the entire observable universe would have been the size of our solar system. Within that crucial first microsecond, all particles had energies higher than can be experimentally achieved in the biggest accelerators on Earth. But that hasn't stopped cosmologists from speculating about an even earlier era, when the entire observable universe was squeezed to a microscopic size. Some physicists hypothesize a primordial phase when our universe emerged from a fluctuation in the quantum haze that existed before $t = 0$. In this quantum haze, a multitude of time and space "bubbles" were rapidly materializing out of the "vacuum" and then annihilating. Some fraction of those nascent universes had the right properties to begin expanding. One of those became our universe.

One could argue that all of these stories are threads in the tapestry of human culture. All express our desire to know and understand and make order. All express our imagination and even artistic sensibility. But the scientific story has a different character. Much of it can be tested against the physical world in addition to the world of our literary imagination, and much of it has indeed been confirmed. In 1929, with our telescopes we observed that the galaxies are flying away from one another, and, in fact, flying away from one another in such a manner as to show that the universe as a whole is expanding. When the rate of expansion is measured, again by scientific instruments, we can conclude that our universe originated in an extremely dense state about 14 billion years ago. (Such accurate dating, of course, is totally impossible in the Sumerian and Chinese origin stories.) Furthermore, our modern theories of quantum physics, confirmed in the laboratory, show that at the subatomic scale, matter and energy can spontaneously appear and then disappear, in what we call "quantum fluctuations." Although we cannot know for sure, an extrapolation of these theories back into the past suggests that our entire observable universe, packed into a region smaller than a single atom, could have been created in a quantum fluctuation.

The scientific story is no less amazing because it is based on experiment and observation, mathematics, and logical reasoning. In fact, it is more amazing because of those foundations. It is astonishing that we six-foot-tall creatures—with a mere three pounds of neurons in our skulls, with our limited range of seeing and feeling, with our relatively short lifespans, living on a speck of a planet in the infinite cosmos—can figure out as much of the universe as we have. And it is all through the activity we call science, the same activity that produced antibiotics and internal combustion machines and computers.

Science and its ideas are part of our culture, our collective imagination and the things we create from it. When the novelist Salman Rushdie spoke at MIT in December 1993 upon acceptance of an honorary professorship, he said that literature thrives on new ideas, and no human activity generates new ideas more richly and rapidly than science. A small sample of those ideas are natural selection and the evolution of species; the idea that all things are composed of units of matter we call atoms; the second law of thermodynamics, that order inevitably yields to disorder; the Big Bang; DNA; the relative nature of time; the cell as the functional unit of life. One of the characters in Vladimir Nabokov's novel *Ada* (1969) makes a specific reference to Einstein's relativity theory. One of the characters in Thomas Pynchon's novel *The Crying of Lot 49* makes a specific reference to the second law of thermodynamics. But these examples are just the tip of the iceberg.

Between 1765 and 1813, a group of scientists, artists, philosophers, and other intellectuals met in Birmingham, England, to discuss the new ideas bubbling up in science and technology, such as the discovery of oxygen, the steam engine, the optical telegraph, and the electrical nature of lightning. They called themselves "The Lunar Society," because they met about once a month, on the night of the full Moon. The extra light made it easier for them to find their way home in the absence of streetlights, probably after considerable inebriants. Members of the group included James Watt, one of the inventors of the steam engine; Erasmus Darwin, an inventor, physician, poet, and grandfather of Charles Darwin; Thomas Day, an author and abolitionist; Joseph Priestley, a chemist and discoverer of oxygen; John Whitehurst, a clockmaker and geologist; and others.

An informal member of the society was painter Joseph Wright (1734–1797), who not only painted portraits of members of the society but also captured the excitement and pathos of scientific discovery. One of Wright's most famous paintings is titled "*A Philosopher giving that lecture on the Orrery, in which a candle is put in place of the Sun*," first exhibited in 1766 and shown below.

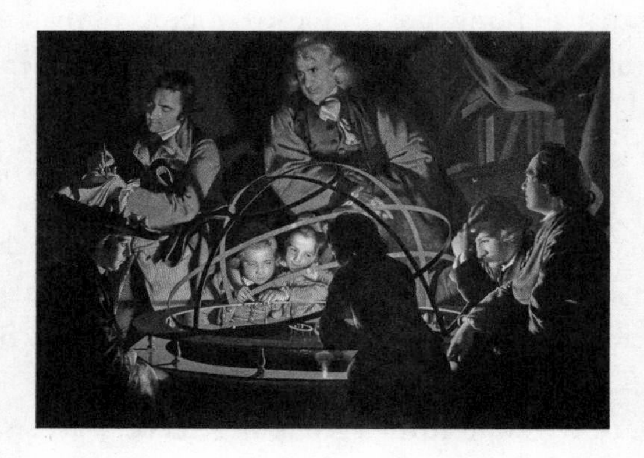

The painting depicts a man giving a lecture about the wonders of the universe to several people, including three children. In the middle is an orrery, a model of the solar system. The children's faces are beautifully illuminated by the unseen candle at the center, standing in for the Sun. As the essayist Sir Richard Steele wrote in 1713, the orrery "administers the Pleasures of Science to anyone." And we can certainly see the pleasures of science in the faces of the children, as well as in the fascination and delight of the other onlookers. Wright seems to have modeled the face of the man on the right from Isaac Newton.

The *Philosopher and the Orrery* is a happy marriage of science with art, youth, education, and human delight. In this painting, Wright has captured our human curiosity, our imagination, and our pleasure in being part of this grand cosmos we find ourselves in.

Science has created our world. Science expresses who we are. Science shapes the future. And its practitioners are prophets of that future, like poets and lawyers and teachers. We are all guardians of the future, and witnesses to this grand performance of a universe.

Chapter III

A DAY IN THE LIFE

In humanizing scientists, we think it might be helpful to meet a few of them. One passionate young researcher we talked to is Lace Riggs. She specializes in the relationship between psychiatric disorders and the detailed activity of brain cells. The study of the brain, or neuroscience, is currently one of the hottest areas of scientific research.

The human brain may be the most complex known object in the universe. It is more complex than thunderstorms, volcanoes, stars, and even entire galaxies. Somehow, in a way still not well understood, the 100 billion neurons of the brain and the million billion filamentary connections between them give rise to love and fear, thinking, poetry, the ability to imagine the future, and that most fundamental and baffling of all mental sensations: consciousness and a sense of self.

Some inkling of the complexity comes from our attempt to duplicate the workings of a brain with advanced computers. The largest computer simulations of brains have about 10 million "digital neurons," fewer than in a mouse. But not only fewer. In these simulations, each neuron is represented as a simple switch—either on or off. By contrast, each of the 100 billion neurons in a human brain is a variable dial. The new large language models of artificial intelligence such as ChatGPT are far faster than human brains in excavating mountains of data. But they lack the ability to ponder, to feel emotion, to have a perspective on the world, to understand context. Indeed, they are probably a long way from anything we would consider consciousness.

A better understanding of the brain will not only advance basic biology and knowledge of that most complex of all objects. It will also help us treat mental disorders such as autism and depression, improve human health and well-being, and strengthen our grasp of who we *Homo sapiens* are.

Although it's hard to get a firm number, there are something like 150,000 neuroscientists worldwide. Four hundred of them work at the McGovern Institute for Brain Research, a part of the Massachusetts Institute of Technology (MIT) in Cambridge, Massachusetts. The McGovern is a microcosm of brain research, with investigations ranging from study of the chemicals that allow neurons to communicate with one another, to using light-sensitive proteins to manipulate the activity of individual neurons, to computer programs that mimic the brain's processing of visual and auditory input, to the study of compounds that might cure mental disabilities.

Lace Riggs is one of the four hundred neuroscientists at the McGovern Institute. As of 2025, Dr. Riggs is in her midthirties.

Dr. Riggs comes from difficult beginnings. She grew up in Southern California's Inland Empire, an area struggling with drug addiction, mental illness, and suicide, all of which engulfed her own family. As she recalls, "Living in severe poverty and relying on government assistance without access to adequate education and resources led everyone I know and love to suffer tremendously, myself included." Lace goes on to say, "Despite everything around me, I had one power, which was the ability to make a decision." In high school, Lace was able to enroll in a special program designed for at-risk students with no financial or social support, enabling her to take courses that counted both for her last two years of high school and first years at a community college. During this period, Lace and her family were moving from house to house, sometimes expelled within a month because her mother could not pay the rent, many of their belongings deposited into storage. She was selected to join the special high-school program because of her answer to an essay question: "How would going to college impact your life?" At Moreno Valley College, Lace received three associate's arts degrees: in social and behavioral science, in humanities and social sciences, and in natural science. After that, she transferred to California State University and eventually earned a full bachelor's in biological psychology.

Initially, Lace thought she would go into psychiatry or become a drug and substance abuse counselor. Then she got interested in the brain. "I realized that some people are more at risk for addiction and mental illness because of their biology," she told us. "I learned that psychiatric conditions are neurologic in origin, the brain being the means through which we perceive and respond to the world, both within and around us."

After shadowing a neuropsychologist for a time, she observed that the available treatments were not good enough. "Instead of being the physician who prescribes only what's been approved by the FDA or the insurance companies, I could make the discoveries myself." Lace went back to school and in 2021 received a PhD in neuroscience from the University of

Maryland School of Medicine. Within two weeks of defending her doctoral thesis, she joined the McGovern as a postdoctoral fellow. Lace Riggs is an example of a scientist motivated from the beginning by a desire to improve human health and well-being—in her case, to help people with mental disorders.

One rainy day in August 2023, we met Dr. Riggs in her laboratory at MIT. The McGovern Institute for Brain Research is a seven-story modern building of white walls and glass. Walking through the huge doors into the lobby, one is treated to a spectacle, a fabulous mobile hung from the ceiling several floors up and consisting of a hundred gold neurons, complete with winding filaments. Viewed from the third-floor atrium, the collection of neurons magically takes on the shape of a human brain.

Dr. Riggs has silky blond hair that falls almost to her waist. We expected that she might be wearing a lab coat, hairnet, and face mask, but she is dressed today in blue jeans, a red sweater, and white sneakers. We had already exchanged a few emails, and she greets us with a warm, radiant smile. Dr. Riggs works in the group of Professor Guoping Feng, associate director of the McGovern. There are about sixty researchers in the group. "I love everyone here," says Lace. "I got so lucky. I have developed a lot of close friendships with people in the lab—people I can be my true self with."

The researchers work in a number of different rooms and wet labs. Lace shares her own lab with two other young scientists, who sit quietly at their computer screens and barely look up when we enter the room. Shelves are laden with glass and plastic bottles of ethanol, phosphate solutions, sodium chloride, and other chemicals; white rubber gloves; narrow glass tubes called pipettes for injecting and withdrawing fluids; and various other items unfamiliar to us. One shelf carries plastic boxes of Lace's personal equipment, with such labels as "Lace Surgery" and "Lace Perfusion." She's meticulous about her tools (scalpels and knives and pipettes) and how they are organized, so that they're kept in small containers within larger containers. "I like to color-code everything," she says, "including my notebooks and folders." In fact, her entire lab is as neat as a bank teller's counter. "But my apartment is a mess," she says with a laugh.

There's some impressive equipment here. On the tables are large digital display screens. One machine that looks like an audio amplifier with all the resistors and capacitors visible on top, called a pipette puller, accurately shapes glass tubes to extremely small points for administering tiny amounts of fluid. There's a centrifuge for separating particles of different sizes and densities, and a high-precision machine called a vibratome, used to cut thin slices of the brains of mice and rats. A refrigerator against the wall holds the tiny intact brains of mice in test tubes filled with formaldehyde. Their brains are not that dissimilar from human brains and are often used by neuroscientists. Lace hands us one of the test tubes. We're both appalled and intrigued by the pea-sized brain floating there. At the moment, the air smells fresh, but Lace says that it reeks with the pungent odor of ethanol and bleach when experiments are underway.

Dr. Riggs goes to her personal desk, near a large window looking out onto the MIT campus. On a shelf above her desk are her color-coded lab notebooks, containing the protocols and results of experiments handwritten in red ink; a photo of a man named Sam, whom Lace describes as "a father figure" to her, about fifty years old in the photo, bald-headed, with a symmetrical goatee and a warm smile; several to-do stickers pinned to the wall; and a graceful green and purple plant, whose tendrils drape down to her desk. "At home, all my plants are dead," she says. Above the desk is a printed statement: "Any pain you feel now—it is nothing compared to what you have already proven you can endure."

One more item over the desk: a lab film strip with the telltale marks showing that particular genes have been disabled, using a gene editing approach known as CRISPR/Cas9. The film shows that CRISPR/Cas9 has unfortunately disabled many genes, rather than the single gene for autism that CRISPR/Cas9 was designed to target and that Lace had intended to study. "That experiment was one of my failures," says Lace. She keeps the evidence in plain view to remind herself that she should not accept received wisdom about the purported action of particular chemicals. "I should have done the validation myself, with my own hands," she says, matter-of-factly.

Dr. Riggs talks about another of her failures, part of the ebb and flow of her career. When she was a graduate student at the University of Maryland School of Medicine, Lace hypothesized that a drug called ketamine, a known antidepressant, could relieve post-traumatic stress disorder (PTSD), a condition she herself has suffered from for years. Her experiment involved fear learning in mice, a process in which a signal, like a musical tone, is always followed by an unpleasant electrical shock. After many repetitions, the mouse expresses fear simply by hearing the tone even with no shock—a model of PTSD. If Lace's hypothesis had been correct, administering ketamine and its derivates would eliminate persistent fear response. But it did not. Lace was so certain of her hypothesis that for an entire year she kept repeating the experiment under different conditions, with the same negative result. "I reluctantly let go of the project, after I had exhausted all reasonable options," she says. "I fell prey mainly to two things: I allowed my personal experiences to skew the objectivity of my scientific judgment. And there's a positive publication bias, in which negative or inconclusive results are not published as often."

While talking, Lace pauses to get a cup of coffee. "The most important piece of equipment in the lab is the coffee machine," she says, smiling. This morning, however, Lace didn't have to wake up so early. But when she has an experiment running, which can take up to ten hours each day, Lace will sometimes get up as early as four thirty a.m. and exercise before walking from her apartment to MIT. (She recently got a bicycle, which may change her mode of transportation.)

Beyond her experiments, Lace has informal meetings with other lab members, especially when they are advising one another on projects and techniques. With her particular background, she consults most often on the signs of psychiatric and autism-related disorders in mouse models. The entire Feng lab meets once a week, during which one or two people present their progress over the last few months. In this way, everyone in the lab knows something about what everyone else is doing.

In a nearby room, Lace showcases her pride and joy, a patch-clamp electrophysiology rig, which she built herself from component parts. She uses the rig for many of her experiments. It's a Rube Goldberg–looking contraption—part microscope, part computer, part camera, and part electronics playground. The rig is able to isolate and manipulate single cells, show them on a digital screen, and record their electrical activity. It costs about $150,000. "It's amazing that the lab provided all of this equipment for me to do my work," says Lace. Taped to the frame of the rig are notes to Lace from multinational visitors and friends, written in Chinese, German, Spanish, and English. "The notes give me good juju," says Lace, "so that I'll have a good patch day." (The "patch" refers to a recording device to measure the electrical activity of cells.)

Lace insists on giving a demonstration. Using the microscope with an attached camera, "I can zero in on a single neuron," about one-thousandth of an inch in diameter. Lace relays a picture of a neuron to the digital screen. There, it's the size of a baseball. She then approaches that single neuron with a very fine glass pipette, the tip of which is far smaller than a neuron. We're watching all of this neurological drama on the computer screen, as if in a miniature movie theater. With another instrument, Lace precisely positions the pipette. She's operating with the skill of a brain surgeon, but she's working at a scale several thousand times smaller than do most hospital doctors. With the microscope, pipette control machinery, and digital display, she very gently moves the pipette tip near the neuron. Then it gets personal. By blowing through a plastic tube attached to the pipette, she clears out all the material around the neuron. Next, she touches the tip of the pipette to the cell wall. Now, she sucks on the tube, causing the pipette tip to bind with the cell wall. With one further puff of her lips, the technical name for which she says is a "kiss," she breaks the cell wall. The contents of the cell flow by capillary action into the pipette. There those contents speak in the language of tiny electrical signals picked up by a wire in the pipette.

"For most people, a neuron is an abstraction," she says. "What exactly is it? When you do this kind of work, come into close contact with a neuron, see it and feel it, move it, listen to its electrical output, you know what it is." She hesitates. "Isn't it f***ing amazing what we know and can do! Doesn't it blow your mind!"

The electrophysiology rig includes Lace's personal vintage oscilloscope, a device made in the 1930s that graphically displays electrical voltages. Compared to the other modern components of the rig, the oscilloscope looks like the antique that it is. Lace explains that she brought it with her when she joined the McGovern, like a plumber taking his favorite, well-worn water-pump pliers everywhere he goes. In most labs, oscilloscopes were replaced by more advanced digital devices in the 1990s. Why do you keep this old machine? we ask. "To remind me where we've come from," she says. "We take a lot for granted now that neuroscience has become so technologically advanced."

A colleague, Tingting Zhou, drops by to say hello. Tingting is also in her early thirties and works on schizophrenia. She's one of Lace's best friends. They go together to jujitsu classes every Friday. Lace describes Tingting as a "badass," and Tingting describes Lace as "very compassionate." It was actually the jujitsu classes, rather than the delicate micro brain surgery, that forced Lace, reluctantly, to cut off her long fingernails. "I prefer long nails," she says, "just like wearing high heels. It's a preferred aesthetic even though they're cumbersome."

Our modern understanding of the brain began with the work of the Spanish neuroscientist Santiago Ramón y Cajal (1852–1934). Before Cajal, it was believed that the nerve cells of the brain, the neurons, formed a single, continuous system, like a large group of people all holding hands. Cajal found that there were tiny gaps between neurons, called synapses. The hands don't quite touch. Neurons communicate with one another by sending chemicals called neurotransmitters across these microscopic gaps.

Each neuron has three parts: a central body; a single long cable, called an axon, several times thinner than a human hair, which sends outgoing information via an electrical impulse; and several projections called dendrites, like tree branches, which receive incoming information from other neurons. When an outgoing message reaches the end of an axon, it stimulates the release of neurotransmitters from other treelike projections. These chemicals migrate across the synaptic gap and are then received by the dendrites of nearby neurons.

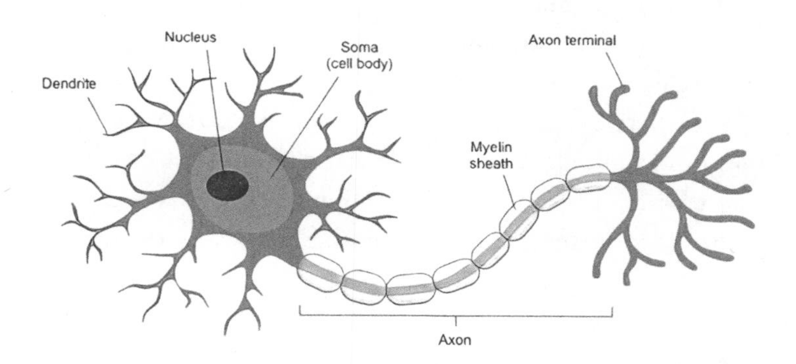

It is the specific network of connections between neurons—which neurons connect to which other neurons—and the amounts and timing of neurotransmitters exchanged between them that regulate everything from memory to thinking to happiness or sadness.

One of Dr. Riggs's recent research papers, with the impossibly arcane title "(2R,6R)-Hydroxynorketamine Rapidly Potentiates Optically-Evoked Schaffer Collateral Synaptic Activity," describes her investigation of a new compound, called HNK, which promotes the release of neurotransmitters that may aid in the treatment of depression. "I was excited," Lace says, recalling the experiment. "It was the first time I had used an optogenetics approach in my work. But a part of me was worried as well. I was worried about whether our hypothesis was true. I knew that this experiment was either going to provide support for what I had already published, or refute it." Lace and her collaborators, Dr. Todd Gould and Dr. Scott Thompson at the University of Maryland School of Medicine, injected into anesthetized rats a special light-sensitive protein that joined the ends of particular axons responsible for releasing glutamate, one of the neurotransmitters. Then, by stimulating those axons, with and without the presence of HNK, Lace and her collaborators succeeded in showing how HNK works, with implications for the development of new drugs.

"I felt very proud," remembers Lace. "All of the first injections I did worked. When you see those electrophysiology tracings, it is the most rewarding feeling. And the fact that the result was consistent with our hypothesis was very reassuring."

Lace shows us one of her living mouse subjects, which has made a nice bed for itself out of cotton and shavings. It scurries about in its cage. The mice are managed by veterinarian technicians, but when Lace is doing an experiment, she personally takes the mice out of their cages, holds them in her hand, and pets them. "We want them to be calm," she says. "I like handling the mice. It does bother me when I have to extract their brains. But we could not otherwise study how the brain works, at least not enough to develop new treatments. I always say 'thank you' to the mouse.

"The public doesn't know what we're doing in science and why we're doing it," says Lace. "What we do is very abstract. It's so far away from everyday life. I think the public's perceptions of scientists are reflected well in the TV series *The Big Bang Theory*. Idiosyncratic and quirky. A lot of people respect physicians. They're the ones who are out there treating our ailments. But who are the people developing those treatments to begin with? It's scientists."

Like many scientists in university settings, Dr. Riggs helps train new scientists. Part of her day includes the mentorship of other postdoctoral researchers, graduate students, and undergraduates in neuroscience. She enjoys these activities. "I train them in experimental procedures. I follow their progress. I also mentor them in more informal ways, like when undergrads go to apply for graduate programs, coaching them on how to deal with the stress of some of the bureaucracy in science and academia."

Our visit today has interrupted Dr. Riggs's usual routine. But on a typical day, she will work until eight p.m. or ten p.m. at night, with the occasional six p.m. to seven p.m. When she gets home, she cooks dinner with her husband, Jason, catches up on leftover email, spends time with her cat, Kori, and then she and her husband watch "mostly mindless but semi-educational YouTube videos" until she falls asleep.

We ask Lace if she works on weekends. "I'm too excited about my work to take weekends off," she says. "What will I do on the weekend that's better than being in the lab? One of the advantages of being in science, especially at the postdoc phase, is that I can determine my own schedule."

Dr. Riggs acknowledges that she does have hobbies, although it's hard to figure out when she has time for them. "I am mostly on the artistic side," she says. "I draw a lot. Music is my other love. Mostly writing music, playing music, listening to music. I considered pursuing music as a career but decided on science instead. Another part: Because of where I grew up, it is very diverse there. I am interested in other cultures and other languages. A lot of the music I listen to is in other languages. I use that as a way to become familiar with how other cultures express themselves."

Lace says that she has few close friends outside the lab. "If I were to paint a picture of who I am now and how I exist in the world, I am more of a lone wolf. I really feel like I am on my own. It's just the way the circumstances are." She does keep up with the people she grew up with in Southern California, but only on Facebook. "No one can believe it, that I could make it from that place to what I am doing now. And they respect it and think I will change the world someday."

Perhaps her current work with autism will change the world. She does worry about whether she can advance from her postdoctoral status to a faculty position at a university. "I have confidence in my abilities," she says. "Given the opportunity, I think I can definitely rise to the occasion. But whether I can convince a faculty hiring committee of that, I don't know."

Lace is playing a jazzy flute piece, "Foolin' Blues," by Lee West. Her instrument, with a gold mouthpiece, she inherited from her grandfather. Seemingly undaunted by an audience, she plays with the same self-confidence and passion as in her science. "I'm a little out of practice," she says. We're in her apartment, on the ninth floor of a newish building about a mile and a half from MIT. It's a two-bedroom apartment that

Lace and her husband can barely afford. "Cambridge is expensive," she says.

When we first come in, her cat, Kori, makes a brief, nervous appearance and then hides under the bed for the rest of our visit and doesn't come out even when her timed food dispenser rings a bell.

Each room of the apartment is small. On the wall of the sitting room, there are a dozen framed photos of Lace and her husband. It's hard not to notice the absence of photos of her family. "I'm not in touch with them," she says. "I don't even know where my sister is." When she speaks of her dysfunctional family, there's a sadness in her voice, but also a resignation. And a reaffirmation of the choices she has made. "Why someone like me with my background gets to this point is because literally I will not give up. I'm very stubborn."

A brown cardboard box of commendations, awards, and diplomas has been left in a corner of her bedroom for a year and a half. "I was intending to hang those," says Lace, "but I never got around to it." On her desk, a fountain pen that she uses to write notes to people "when the words should have some weight." Another item on her desk: a white disk that glows gently when turned on. "It's a SAD lamp a lab friend gave me," she says. "Seasonal affective disorder. I use it for putting on makeup. I have to make my eyebrows from scratch, you know."

Lace's worlds merge in this space nine floors up. In the sitting room, a bookcase contains titles ranging from fantasy novels by Kevin Hearne to Tolkien's *Lord of the Rings* to Plato's *Republic* to B. F. Skinner's *About Behaviorism* to textbooks on biology and neuroscience. On a shelf rests an old manual Royal typewriter. "I use it to write poetry," she says. In the small kitchen, six bottles of pinot noir, unopened. In the bedroom, a microscope on the dresser, a rumpled bed.

"Would you like something to drink?" she asks after her performance on the flute. She makes coffee for herself. She offers to walk us back to our car.

Lace doesn't know how long she'll be able to stay in this apartment, or how long she'll keep working at the McGovern before applying for faculty positions across the country. But she does know what she wants to do with her life. "My main motivation now is to make a difference. It may not be immediate. The research may just sit in a paper in PubMed for twenty years, but eventually if that contributes something to improve our understanding of the brain on the experience side, to me that is the most rewarding thing that we can do."

Chapter IV

HOW SCIENTISTS THINK

The landmark sci-fi film *The Matrix* (1999) begins at some point in the future when all human beings experience a totally simulated reality, their minds controlled from birth by malevolent machines. It takes quite some time for Neo, the main character, to discover that the world he lives in is a complete illusion and that he can access the "real" world only with great difficulty.

The nature of reality has long been a subject of fascination not only for filmmakers, but also for philosophers, novelists, and scientists.

Scientific thinking is founded on two key assumptions, often not stated explicitly but present in the very air scientists breathe. First, that an external world exists outside of our minds. And second, that external world is lawful. These two assumptions are not as simple as they seem.

Everything we know of the world—every sight and sound and thought—is necessarily processed and filtered through our minds. Even when we look at the position of the stars overhead or check the reading of a voltmeter, we are considering that information with our minds. The mind, of course, can create its own version of reality. So it might be argued that there is no external reality at all; there is only mind and perception. Exactly this point of view has been suggested by some distinguished philosophers. And one of the world's greatest religious traditions, Buddhism, advocates the idea that tables and chairs, although "real," do not have an intrinsic nature independent of our perceptions of them and their connections to everything else in the world.

Scientists do indeed have mental preconceptions about the world, psychological and philosophical prejudices. But, despite these mental projections, we scientists do believe that stars and mountains and rainbows exist whether or not we see them or touch them or imagine them. One argument in favor of the existence of external reality is that scientists are sometimes surprised by what they discover, overthrowing previous beliefs. As mentioned in chapter 1, examples include the miasma theory that all diseases are caused by bad air rising from the ground, the idea that heat is a fluid, the theory that miniature human beings reside whole inside eggs and sperm, and the belief that a very tenuous and transparent substance called the aether fills up space and is responsible for the propagation of light. Of course, one could argue that these "surprises" are also an illusion, a mental construction. But that view would require an additional layer of mental fabrication. Given all of human experience through the ages, not only in science, it seems far more likely that some external reality exists beyond our minds, rather than that we are all living in a *Matrix*-like world.

The second key assumption of science, also with some subtlety, is that the physical world is lawful. That is, each event, large or small, was caused by some prior event, and those causes obey rational and logical rules. The rules do not act at some times and not others, or at some places and not others. The rules are always and everywhere the same. An example of such a rule: an object in motion does not change speed or direction unless it is acted upon by a force. Every scientist, consciously or unconsciously, works within a framework built with this assumption. Without a lawful world, wheelbarrows could suddenly begin floating for no reason; the Sun could turn on and off at random; people could change into salamanders. Scientists spend their time discovering, observing, and testing the laws of nature. But if there are no laws, or if the laws are violated now and then, the scientific enterprise is futile, like building a sandcastle on the edge of the ocean, with waves continually surging in to wash it away.

There are some complications to this seemingly simple assumption of lawfulness. First is the question of whether there is a nonphysical universe in addition to the physical universe, not subject to laws—a belief in fact shared by most of the world's population. Scientists require either that such a nonphysical universe not exist, or that the two universes never interact with each other. But then the definition of the physical universe ends up with an empty tautology: the physical universe is everything that is subject to the laws of nature. Secondly, if the universe is lawful, as scientists assume, are there final forms of the laws, needing no further approximation? The late Nobel Prize–winning physicist Steven Weinberg wrote a book titled *Dreams of a Final Theory* in which he argues for such ultimate laws. As we will discuss below, science is a provisional activity, in which we are constantly revising our theories as new data and ideas become available. In fact, the history of science may be viewed as a project making better and better approximations of the way that nature behaves. Will this process of revision continue indefinitely? Some scientists believe that final laws don't exist. And if they do exist, are we human beings capable of discovering them? Even if final laws do exist and we discover them, we would never know with certainty that we were in possession of such final laws, because we could never be sure that some new phenomenon tomorrow might fall outside of the currently accepted laws and require their revision.

In practice, scientists do not fret much about these complications. At an early stage of our apprenticeship, we scientists begin assuming that an external reality exists and that it obeys laws. Other than these two fundamental assumptions, science is not an ideology or a belief system. Science is not a collection of facts. Science is a way of thinking. It is a way of investigating the physical world, attempting to organize the behavior of the things observed, searching for patterns and rules that govern those things, testing the hypothesized rules, and then revising the rules when they disagree with an experiment. As mentioned in chapter 1, this manner of investigating the physical world, often called "the scientific method," might better be called "critical thinking," and it is not owned by scientists. Many other kinds of professionals, such as lawyers, doctors, auto mechanics, and accountants, use evidence-based critical thinking in their work.

As mentioned in chapter 1, the person who is often credited with being the first to systematically put the scientific method into practice was the eleventh-century Egyptian physicist Alhazen (Ibn al-Haytham). Most scholars, perhaps with a Western bias, trace the beginnings of modern science, the so-called Scientific Revolution, to Europe in the seventeenth century.

At that place and time, the natural philosophers of the day (who were not called scientists until the nineteenth century) began to base their theories on observations of nature rather than on received wisdom or untested conjectures. One of the first people to advocate for this approach was the English philosopher and statesman Francis Bacon (1561–1626). The Italian physicist Galileo put Bacon's ideas into practice by doing such experiments as carefully measuring the time it took for balls to roll down inclined planes and then formulating his laws of falling bodies. Around the same time, the German astronomer and mathematician Johannes Kepler (1571–1630) analyzed the scrupulous astronomical observations of his boss, the Danish astronomer Tycho Brahe (1546–1601), and found patterns in the orbits of planets, such as that the squares of the orbital periods of the planets are directly proportional to the cubes of the semi-major axes of their orbits (now known as Kepler's third law). In simpler language, the time for a planet to orbit the Sun, multiplied by itself (i.e., squared), equals the distance of the planet from the Sun, multiplied by itself twice (i.e., cubed).

All of these scientific investigations culminated in the magisterial work of Isaac Newton, his laws of motion, his law for gravity, and many other discoveries and formulations of laws of nature. Besides being a mathematical genius, Newton actually did experiments and tested his hypotheses whenever possible.

Here's a simple activity involving pendulums you can do yourself that illustrates many aspects of the scientific method/critical thinking at work. First, some terminology. The time for a pendulum to make one complete swing back and forth is called its period, and the mass at the end of the pendulum is called its bob. Two obvious scientific questions are: (1) How does the period depend on the weight of the bob? Does the period increase or decrease with increasing weight? (2) How does the period of the pendulum depend on its length?

Now, let's make some hypotheses: (1) It is reasonable that the period should be *shorter* for heavier bobs, since gravity is stronger for heavier objects, pulling a heavier bob back to the vertical position more quickly. (2) It is reasonable that the period should be *longer* for pendulums of greater length, since the bob has farther to swing in each back and forth.

If you were Aristotle, at this point you would put down your reed and jar of octopus ink, roll up your papyrus, and consider yourself done. But scientists test their hypotheses.

You can construct a pendulum by tying a paper clip to the end of a string and then hanging keys from the paper clip. The keys are the bob. To test hypothesis (1), you can make a pendulum with one key at the end, set it swinging, and time its period with a stopwatch. Then add a key, making the bob heavier, and time the swing. Then add a third key, making the bob even heavier. To get an accurate period, since there might be a bit of variation in how you release the pendulum and small mistakes in the starting and stopping of your stopwatch, let each pendulum make four complete swings and then divide that total time by four. All experiments have a bit of experimental error, and the scientist needs to compensate for these errors if possible—which the averaging here does. Possibly to your surprise, you will find that the period is the same no matter how many keys you put on the paper clip. You have disproved the hypothesis. The period of a pendulum does not depend on the weight of its bob. You might be very fond of your hypothesis, since it makes perfect sense, but you will have to give it up if you are acting like a scientist.

To test hypothesis (2), you need to make pendulums of different lengths—which you can do by using strings of different lengths. For each new pendulum, measure the length of the string and then time the duration of the period. You should find that your hypothesis (2) is correct. The period for longer pendulums is indeed longer.

But you can do much more with your results. If you make a graph of the period versus the length of your pendulums, you will get something like this:

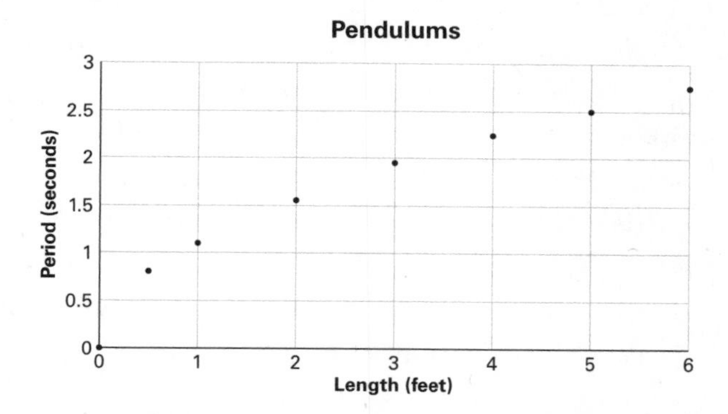

Each dot on the graph comes from the period and length of a different pendulum. Each dot is a different data point. If you draw a curve through your data points, it looks like this:

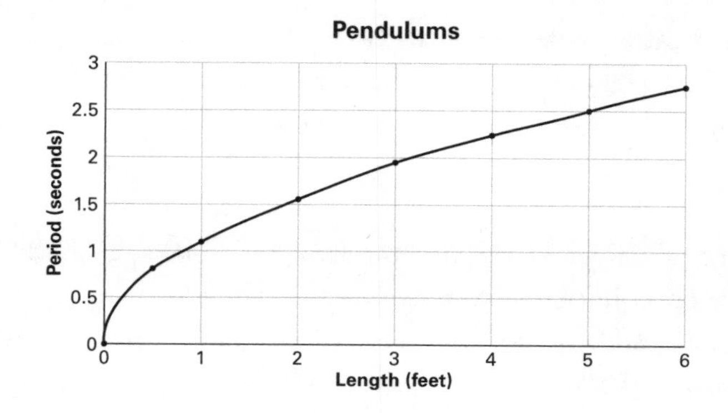

The more data points you have, the smoother the curve. In fact, you can now use this curve to *predict* the period of a pendulum even before you have made it and timed it. Try that out. Make a new pendulum at a length you have not made before, use the curve to predict its period *before letting it swing,* then measure its period. Voilà! You can make a successful prediction with the law you have found for pendulums. The curve is the law. This particular curve is called a parabola. You have found a universal pattern, or law, for pendulums. You have ruled out your first hypothesis. And for your second hypothesis, you have not only confirmed it, but you have found a quantitative relationship between the period and the length of a pendulum. And it is a universal law, applying to all pendulums.

Mathematically, the equation for the pendulum law you have found is:

$$\text{Period} = 1.1 \text{ seconds} \times \sqrt{\text{length (in feet)}}$$

where √ is the symbol for square root and can be found on most calculators. With some calculus and Newton's laws, you can actually *derive* the pendulum law and relate the 1.1 seconds in the law to the strength of gravity on the Earth. For the Moon, which has less gravity, the number would be 2.7 seconds. But even without going that far, you have used the scientific method to test your two hypotheses.

Science is always a work in progress. All scientific theories should be considered as provisional. They must be revised when new information or experimental results conflict with the existing theory. And, in fact, that is the history of science.

Some theories, like the germ theory of disease, are supported by overwhelming evidence; others, like the idea that vitamin C helps prevent colds, are more tentative, awaiting further experimental confirmation or refutation. But however confident scientists may be in a theory, they should keep their minds open, or at least ajar, to the possibility that some intellectual revolution will offer a drastically different perspective. The revolutions in science, described by the late American philosopher of science Thomas Kuhn, are actually very rare. Still relevant is Francis Bacon's injunction: "If a man will begin with certainties, he shall end in doubts; but if he will be content to begin with doubts, he shall end in certainties."

We should say a few words about what constitutes a theory in science. A scientific theory is much more than a conjecture or a guess. A scientific theory is a foundational principle or mathematical system that is either based on direct experimental evidence or that logically follows from other principles and systems based on evidence. For example, Darwin's theory of natural selection is based on fossils and the similarities and differences between living organisms. String theory in physics, which unifies all the fundamental forces of nature, is not directly based on experimental evidence, but it is supported by underlying and evidence-based quantum physics and relativity physics, and it follows a pattern of other theories in physics that have successfully managed to unify some but not all of the forces of nature. (It is important to point out that string theory has not yet been experimentally tested, so it is much more tenuous than Einstein's theory of relativity, for example.) The statement that all living organisms in the universe are based on carbon chemistry would be a conjecture but not a scientific theory, since it is obviously not based on any direct evidence and does not necessarily follow from the limited evidence we have on our own planet. Other life-forms could be based on silicon, for example.

Scientific theories are generally more than single statements or equations. They are large conceptual frameworks that apply to many different phenomena.

Scientific theories also make predictions, and they should be testable. If a theory is not testable, it might be very interesting, but it does not belong in the domain of science, and graduate students in physics or biology or chemistry would do well to steer clear of such a "theory." Scientific theories can be disproved, but they can never be proved, since we can never be sure that another piece of data discovered tomorrow might not violate the theory. Sometimes, there is no clear distinction between a scientific theory and a single law of nature, such as the second law of thermodynamics—that all isolated systems evolve from order to disorder—although, as we have said, a scientific theory usually includes an entire intellectual framework for understanding a set of physical problems.

The forward movement of science may be considered as better and better approximations of the behavior of nature. For example, Galileo's law of falling bodies (ca 1590) was given firm foundational footing by Newton's theory of gravity and laws of motion in 1687. For two centuries, Newton's law of gravitation worked beautifully. But in the mid-nineteenth century, with increased precision of telescopes and careful measurements, astronomers concluded that the orbit of the planet Mercury didn't quite match the predictions of the law. The accumulating discrepancy was extraordinarily tiny, about one-hundredth of one angular degree *every century.* However, Newton's law was so well-defined, and the measurements of Mercury so precise, that some scientists were troubled. Then, in 1915, Albert Einstein proposed a new theory of gravity called general relativity.

Einstein was struck by the fact that bodies falling due to gravitational attraction follow the same trajectory no matter what the bodies are made of (a version of the pendulum result). From this hint, he conjectured that gravity did not interact with the particular forces within particular bodies but, instead, interacted on space itself, altering the geometry of space to produce universal trajectories. Then he developed the mathematics to express that insight. Einstein's theory completely explained the orbit of Mercury. Furthermore, it predicted many new phenomena, such as black holes and gravitational waves. Now the main point: Despite its subtlety and enormous success, we know that Einstein's theory will also need revision. It does not incorporate quantum physics, which we believe must be a part of every theory in physics.

Another example of the typical progress of science is the theory of disease. As far back as the fourth century BC, people believed that disease was caused by so-called miasma, bad air emanating from rotting material and swamps. Ideas began to change with the invention of the microscope and the discovery of microorganisms, directly observed and described by the Dutch scientist Antonie van Leeuwenhoek and discussed in chapter 1. Even earlier, and also using the newly invented microscope, the German Jesuit priest and scholar Athanasius Kircher (1602–1680) saw tiny microorganisms, which he called worms, in decaying bodies and suggested that disease was caused by invisible living organisms. In 1700, French physician Nicolas Andry (1658–1742) argued that microorganisms were responsible for smallpox and other diseases. Finally, in the late nineteenth century, the German microbiologist Robert Koch (1843–1910) formulated the modern theory of germs and disease: each infectious disease is caused by one specific microorganism.

The provisional nature of scientific theories—the way that scientists revise their beliefs when new data becomes available—is not sufficiently appreciated by the public, sometimes leading to a mistrust of science and its practitioners.

The public is especially sensitive to changes in scientific opinion in the medical and biological fields. For example, early in the COVID-19 pandemic, the Centers for Disease Control and Prevention (CDC) and the surgeon general, both in the United States, and the World Health Organization all recommended against wearing masks to prevent the spread of the disease. These negative guidelines were based on several factors: most importantly, the lack of sufficient data; the belief that the disease was not widespread; the legitimate concern that the limited supply of masks should be saved for health care workers; and the consideration that, culturally, most Westerners were not prepared to wear masks. However, by 2021 and 2022, the recommendations changed, due to more data. Studies published by the U.S. National Academy of Sciences, the Royal Society in the UK, and by the World Health Organization itself attested to the efficacy of masks during the pandemic. Or consider the changes in recommendations about fat consumption, or vitamins, or a host of other food and medical issues.

Another example of changing medical and health recommendations is alcohol consumption. Until a few years ago, it was commonly stated by the medical establishment that a small amount of alcohol, especially red wine, was actually good for you. However, a January 2023 publication by the World Health Organization, based on a review of a dozen recent studies, stated that "no level of alcohol consumption is safe for our health." It remains to be seen whether this recent recommendation holds up after further study and analysis. (See chapter 8 for a discussion of the responsibility of scientists and science journalists in evaluating and reporting new scientific results.)

Much of the public sometimes seems to discount science altogether because of these changes of position, failing to understand that such changes are part of an evidence-based enterprise. When new evidence becomes available, we should want scientists to be able to change their minds, just as you would want a physician to change your medications if they discover they have misdiagnosed your illness. The important point is that science cannot be dismissed simply because it sometimes "changes its mind" about particular phenomena. This is the way that critical thinking proceeds. At any given moment, scientists try to develop theories and recommendations that are based on the best available evidence at that time. It is the responsibility of the public to understand how critical thinking works, and it is the responsibility of scientists to help educate the public on the nature of science and to use caution when stating their theories and recommendations, not overstating the evidence.

The scientific method, as we have described it, seems the paragon of objectivity. And it is, with the help of the *community* of scientists critiquing one another's work. Scientists tend to be severe critics of other scientists' work, demanding that experiments be reproducible and that the derivation of equations from starting assumptions be rederived by other scientists. But individual scientists are often not objective; they are often subject to the same emotional forces, prejudices, egos, and personal preferences as nonscientists. And when scientists succumb to such emotions and prejudices in their scientific work, they lose the objectivity that is the hallmark of science. As Francis Bacon wrote four centuries ago, "The human understanding resembles not a dry light, but admits a tincture of the will and passions, which generate their own system accordingly; for man always believes more readily that which he prefers."

A twentieth-century example of the loss of objectivity was the announcement in 1969 by American physicist Joseph Weber (1919–2000) that he had detected gravitational waves—fluctuations in the geometry of space caused by cataclysmic events, like colliding black holes. Weber claimed to have detected these waves via vibrations they produced in aluminum cylinders in his lab. Over the next few years following Weber's claims, other scientists built very similar detectors, or even better versions, but could not reproduce Weber's results. Furthermore, theoretical astrophysicists pointed out that Weber's cylinders did not have nearly the sensitivity to detect likely sources of gravitational waves and their intensities. Yet in spite of the contradicting research, Weber continued to insist that he had indeed detected gravitational waves and that everyone else was wrong. Weber was not a crackpot. He was a distinguished physicist and had earlier developed the correct idea for lasers, for which he was never given credit. But with his work on gravitational waves, he lost his objectivity.

Ideally, individual scientists should not become too enamored of their pet theories and experiments nor influenced by wishful thinking. But that is sometimes a tall order. A person who has invested years of their life in a project is bound to be strongly committed to its importance, to the extent that it is a traumatic wrench, as well as a blow to the ego, if the whole effort comes to nothing.

Sometimes, reason and objectivity can go out the window when they clash with preconceived notions. An example can be found in a paper written by Soviet physicist Lev Davidovich Landau (1908–1968), winner of the 1962 Nobel Prize for major contributions to superconductivity, among other things. In fact, Landau was feared by colleagues, partly for his habit of ruthlessly ferreting out and destroying all unproven statements in scientific discussions. The nameplate on his office door at the Ukrainian Physico-Technical Institute read: "L. Landau. Beware, he bites."

In 1932, with several important pieces of work under his belt, Landau published a curious three-page paper titled "On the Theory of Stars." What is shocking about the 1932 paper is that Landau, without warning and in a single sentence, dismisses a major branch of physics. The paper concerns a theoretical investigation of the structure of stars required by balancing their inward gravitational forces against their outward pressure forces. For the burned-out stars Landau was considering, the outward pressure forces are prescribed by quantum mechanics, the theory of matter at the atomic level. Despite being a new field of physics, developed in 1925 and 1926 (see the profile of Heisenberg in this book), quantum mechanics had already proven itself extremely successful in describing atomic phenomena. Quantum physics ranked beside Einstein's relativity as the foundation for modern physics. To Landau's dismay, his calculations predicted that burned-out stars cannot avoid complete inward collapse if slightly more massive than the Sun. (Such a collapse would cause the star to contract from a sphere a million miles or more in diameter down to a point, later called a black hole.) Landau then wrote, "As in reality such masses exist quietly as stars and do not show any such ridiculous tendencies, we must conclude that . . . *the laws of quantum mechanics (and therefore the laws of quantum statistics) are violated.*" Landau had little justification for this statement. Astronomers had indeed observed massive stars quietly avoiding collapse, but they were not the burned-out, cold stars that Landau's calculations applied to. Hot and cold stars were easily distinguished by their colors. It seems that Landau found his theoretical result so disturbing to his own common sense that he was willing to abandon the celebrated theory that produced the result.

Even Einstein was subject to philosophical prejudice. He refused to accept the new quantum physics. Einstein disliked the element of probability (rather than the certainty in all previous areas of physics) that is inherent in quantum physics.

Objectivity and the banishment of emotion are indeed part of the scientific enterprise. But they arise from the society of scientists, reviewing and criticizing one another's work, demanding that all claims be tested and all experiments be reproduced by other scientists. The Weber case is a good example of that process, with Weber's claims eventually being dismissed by other scientists. Another example can be found in the claim in 1989 by two chemists, Stanley Pons and Martin Fleischmann, that they had built a device capable of producing nuclear fusion at room temperature, immediately dubbed "cold fusion" by the press. If their claim had been true, it would have led to a vast and cheap new source of energy. Pons and Fleischmann and the University of Utah, where they had been working, announced the result at a press conference. The press went wild with the possibilities of a monumental new source of energy for an energy-starved world. For the next year, Pons and Fleischmann were international celebrities.

Of course, many other groups of scientists immediately set about to build equipment like that of Pons and Fleischmann. But none of them were able to conclusively reproduce the claimed results. (Some groups initially seemed to confirm the Pons and Fleischmann claims but then found errors and retracted their results.) Within a couple of years, the claim was dead, although a small number of physicists still take cold fusion seriously and hold conferences. This case illustrates both the mistakes that can be made by the unbridled excitement of individual scientists as well as the power and validity of the scientific method as practiced by the community of scientists.

Interestingly, emotion in science is not always a bad thing. It can be the positive force that drives a scientist on, pushing her to work through the long hours of the night, pushing her to spend years on a problem, as was the case with Marie Curie, discussed in chapter 1, or with Barbara McClintock, discussed later in this book. Personal commitment to a scientific problem and the emotion that goes along with it are often a necessary part of the psyche of scientists. As the British-Hungarian chemist and philosopher Michael Polanyi (1891–1976) wrote in his book *Personal Knowledge,* "The distinctive ability of a scientific discoverer lies in . . . his originality. Originality entails a distinctively personal initiative and is invariably impassioned, sometimes to the point of obsessiveness. From the first intimation of a hidden problem and throughout its pursuit to the point of its solution, the process of discovery is guided by a personal vision and sustained by a personal conviction."

Emotion in science is a double-edged sword. It can be an impediment, as in the case of Joseph Weber, or it can enhance engagement when the individual scientist is able to retain some degree of objectivity. Although the gravitational wave detector built by Weber was far too rudimentary to make a real detection, more advanced gravitational wave detectors did indeed eventually succeed. But it took a huge amount of technology to get there and many years of hard work. Physicist Kip Thorne, who shared the 2017 Nobel Prize for successfully detecting gravitational waves in 2015, worked for forty years on the project. What kept him going for that long? One element was emotional involvement and personal commitment. As Thorne said in an interview, "Certainly a passion for what I'm working on has been a big factor in my career. I've chosen to work on things because I was passionate about them. And the passion does play a role in my working very hard over a long period of time. But the passion itself is driven by the joy of the process, and in part by the belief that the payoff is going to be huge."

Even though emotional commitment and ego are embedded in the psyche of individual scientists, their enterprise is modest. For the most part, science advances step by step. If you ask scientists themselves what they are working on, you will seldom get a grandiose reply like "seeking to cure cancer" or "understanding the universe." Rather, they will focus on a tiny piece of the puzzle and tackle something that seems tractable. Scientists are not thereby ducking the big problems but instead judging that an oblique approach can often pay off best. A frontal attack on a grand challenge may, in fact, be premature. To take a historic example: Fifty years ago President Richard Nixon declared a "war on cancer," envisaging such a war as a national goal modeled on the then recent Apollo Moon-landing program. But there was a crucial difference. The science underpinning Apollo—rocketry and celestial mechanics—was already understood, so that when funds gushed at NASA, the Moon landings became reality. But in the case of cancer, the scientists knew too little to be able to target their efforts effectively.

In any case, scientific progress is incremental, and scientists understand this fact of life, even the bold and egotistical ones like Ernest Rutherford and James Watson. And some scientists are truly humble. In his autobiography, Charles Darwin wrote, "I have no great quickness of apprehension or wit which is so remarkable in some clever man, for instance Huxley. . . . I have a fair share of invention and of common sense or judgment, such as every fairly successful lawyer or doctor must have, but not I believe, in any higher degree."

It is sometimes said that the sciences are objective, and the arts subjective. Is this statement true, and if so, what does it mean? As we have argued, the objectivity of the sciences arises from the community of scientists, not from individual scientists.

Science, other than psychology, concerns itself with the external world, the world outside of our minds. Even brain science treats the brain as a physical object, its neurons and synapses to be prodded and explored as an electrician tests the wiring of a household appliance. Although the work of science is performed by individual human beings, the successful experiment is one that can be reproduced in the laboratories of many scientists; the important equation is one that can be rederived by theoretical scientists anywhere. The "product" thus becomes universalized; its meaning and functionality transcend its original creators even if their names remain attached; the scientific product is absorbed into the general body of knowledge about the physical world. In this process, the individuality of the original creators becomes irrelevant. Their fingerprints are dusted off the final piece of equipment or theory or equation. Darwin's theory of natural selection is a universal principle. In no way does it embody the personality or presence or expression of Darwin. Likewise, Einstein's theory of relativity. Or Fleming's discovery of the first antibiotic. The endpoint of science is not personal.

By contrast, in the arts the personal stamp and individuality of the artist are everything. The arts, ultimately, concern the inner world of the artist, the artist's individual reaction to the world and the expression of that reaction. The work of the artist *is* the artist, in a way that natural selection is not Darwin and relativity is not Einstein. If Einstein had not proposed his theory of special relativity in 1905, it is almost certain that another scientist would have proposed the same theory, with the same exact equations, within a decade. But if Beethoven had not composed his Ninth Symphony, it would never have been composed by anyone else. If Gabriel García Márquez had not written *One Hundred Years of Solitude,* it never would have been written by anyone else. (Of course, we can never say never, but the probability of such a duplication is incredibly small, a decimal point with many zeroes after it.)

There's a fascinating paradox here. The scientist, dealing with the external world as she does, surrendering her personal stamp on her product as she does, achieves a kind of permanence. Her work becomes an essential pillar of the scientific world. The artist's work, dealing with the internal world as it does, intimately tied to her personal expression as it is, may be an important page in the history of art, but it is not essential for the work of future artists. Einstein's theory of special relativity and Picasso's cubism were conceived within a few years of each other. Today, more than a century later, a young artist has the choice to include cubist elements in her paintings or not. But a young physicist proposing a new theory of matter must make sure that the theory is compatible with relativity, unless describing the very tiny and extremely dense conditions where quantum physics must be taken into account. Because it is now known from numerous experiments that the external physical world obeys the rules of relativity to a very high degree of accuracy, and a new theory in conflict with relativity is dead on arrival.

Other differences: At any given moment, every scientist is working on a problem that she or he thinks has a definite answer, even if it may take decades to find the answer. In the arts, however, many interesting questions don't have answers, or cannot be phrased as questions to begin with. Why did Raskolnikov kill the old lady in Dostoyevsky's novel *Crime and Punishment*? Does van Gogh's painting *Starry Night* even pose a question?

Despite these differences, science and art both reflect the searching and passion and imagination of us human beings. We need the objective, and we need the subjective. We need questions with answers, and we also need questions without answers.

DOROTA GRABOWSKA

"You go down there and, you know, as someone who has been in many cathedrals in Europe and in churches in the States, it has a very similar feeling because it's this massive expanse, and it's filled with hours and hours and hours of human time and thought and contributions. And that's what a cathedral is." Physicist Dorota Grabowska is describing their place of work, the atom smasher called CERN, the largest scientific facility in the world, located on the Franco-Swiss border near Geneva.

It's a cool day in October 2021, and we're sitting at an outdoor cafeteria on the CERN campus. Occasionally, other scientists and engineers walk by, excitedly talking about bosons and beams. Three hundred feet below us, the CERN's underground tunnel silently winds around in a circular ring 17 miles long. The ring accelerates subatomic particles up to 99.999999 percent the speed of light. Then the minuscule bits of matter collide with one another, producing exotic new particles, some of which disintegrate in 0.0000000000000000000001 seconds—long enough for physicists like Dorota to test their theories of the fundamental particles and forces of nature.

We wanted to talk to Dorota, in part, to find out what it's

like to do science in such a huge place, surrounded by giant machines and hundreds of other scientists. And also to better understand and portray what theoretical scientists do.

Dorota prefers the personal pronoun "they." At the moment, their hair, some of it colored purple, is swept off to one side, framing rosy cheeks and a full face. Dorota is wearing what looks like a ski jacket. Growing up in Colorado, Dorota skied a lot. In fact, they remember always counting the number of chairs on each ski lift, an early sign of their interest in numbers and mathematics.

Dorota is thirty-three years old, an age when many theoretical physicists see their best work in the rearview mirror. It's a young person's game. The discipline requires an almost athletic limberness of mind and a high degree of skill in mathematics. The external equipment: pencil and paper and sometimes computers.

Theoretical scientists make mathematical models of the physical world. As an extremely simple example, a theoretical physicist might model a pendulum as a ball on a string, swinging in a vacuum, and then derive the equation for such an idealized pendulum:

$$l\, d^2\theta \,/\, dt^2 = -g \sin \theta$$

The physicist then solves that equation for various initial angles θ of the release of the pendulum. Finally, the scientist must see how well those solutions match up with real pendulums in the real world. The above equation shows the universality of the laws of nature. The letter "g" stands for the acceleration of gravity. On Earth that acceleration is 32 feet per second per second. But the equation applies equally well on the Moon, where the acceleration of gravity is about 5.3 feet per second per second. Pendulums on the Moon take longer to complete a full swing than on the Earth, all quantitatively predicted by the equation. (See the exercise on pendulums in chapter 4.)

Theoretical scientists spend hours each day working with such equations. One requirement of theoretical scientists, and especially theoretical physicists, is the ability to mentally live in a highly abstract world, where space and time are not what they appear to the senses. Albert Einstein, perhaps the greatest theoretical physicist of all time along with Isaac Newton, once said that "the whole of science is nothing more than a refinement of everyday thinking." Of course, that depends on how you define "refinement."

On a practical level, theoretical physics is a highly competitive field, with its practitioners struggling to do something new and significant. The vast majority of research papers are forgotten as quickly as fast-food hamburgers. And jobs are scarce. Within this environment, Dorota Grabowska works in an even more competitive and rarefied subspecialty, something called lattice gauge theory. Modern theories of physics, in which the positions of particles are only probabilities rather than certainties, describe an infinite number of possible positions and paths from A to B, making some calculations almost impossible. Lattice gauge theory approximates space as a finite series of separated points, the so-called lattice, so that there are only a limited number of points and paths from A to B. Dorota and colleagues have used lattice gauge theory to calculate the way that groups of subatomic particles interact with one another under the strong nuclear force.

The fierce competition, plus the highly abstract nature of

the subject itself, can be psychologically rough, especially since theorists often work alone at their desks. "The highs are high, but the lows are very low," says Dorota. "And so it can be hard to maintain that confidence in your own abilities, when some days nothing goes right, and you can't figure anything out. You get stuck on a problem."

Dorota has been a researcher at CERN for two years. Before that, they did research at the Berkeley Center for Theoretical Physics and the Lawrence Berkeley National Laboratory.

CERN, the acronym for "Conseil Européen pour la Recherche Nucléaire," is a whale of a place. Its largest instrument, ATLAS, which detects the debris from subatomic collisions, is a cylinder some ten stories tall, weighing 7,700 tons, similar to the weight of the Eiffel Tower. Many of the cavernous rooms at CERN resemble futuristic sci-fi scenes of intricate electronics, wires, and tubes. The giant control room, with about fifty workstations, has sixteen 46-inch digital monitors around the walls, keeping tabs on the light-speed beam and detectors. Superconducting magnets, used to bend the whizzing subatomic particles into circular trajectories, are a hundred thousand times as strong as the Earth's magnetic field, strong enough to totally ruin metallic watches and credit cards and accelerate a carelessly placed screwdriver up to a speed of 150 miles per hour. (There are plenty of danger signs at CERN.) The annual budget of the facility: about $1.5 billion.

Something like 12,200 scientists from seventy countries use the data gathered at CERN. On site, the staff includes about 2,500 people—physicists, engineers, IT specialists, human resources specialists, accountants, writers, and technicians.

When asked what it's like to work here, Dorota says that most theoretical physicists work at universities, rather than large transnational labs like CERN, and sit quietly at a desk tucked away in the corner of a physics department. "You make friends outside of the theorists, like the people in the cryo team, in charge of keeping everything cold. People who work on the beam acceleration. It does change the nature of your day when you're surrounded by that, rather than at a university." CERN is certainly not a university. There are no students, no formal classes or grades, no history or philosophy or literature. It's physics, 24/7.

Being a theorist, Grabowska collaborates with only one or two other people on each research project. But the collaborations of experimental particle physicists, the majority of people at CERN, often involve several hundred people, and the list of authors on published papers sometimes covers most of the first page.

It's worth pondering whether the ambition of young scientists, wanting to make their mark, might be thwarted by being only one cog on such a large wheel. "I think that the underlying drive is the same," says Dorota. "I see something when I'm working on a physics problem. 'Ooh, that's shiny, that's interesting, I'm curious about what's there.' And I think that the experimentalists have similar things, where they see something, and it becomes interesting. And now is the question, how does that link into the larger experiment? You still need that same underlying curiosity and interest. But you have to be comfortable with working towards the larger goal."

Grabowska was born in Boulder, Colorado, six months after their parents immigrated from Poland. Dorota is fluent in Polish. Science was always part of their life, starting in childhood.

They grew up in a tight-knit Polish community, in which most of the men had PhDs in STEM, and most of the women had master's degrees in humanities or STEM. Dorota's father is an atmospheric physicist. "We didn't call it science," Dorota says. "It was just a day-to-day part of my life. I noticed the leaves going around corners. The swirling vortices. And then trying to replicate it, trying to see what was happening. While cooking, I would see water levitate on a pan and try to understand that. It's weird to try to identify that scientific curiosity when it is innately a part of you. I loved LEGOs and things like that." Like many budding scientists, and especially those who become theorists, Dorota was always good at math. Later, Dorota realized that some people were much better at pure mathematics, without physical applications. "There are certain ways that my brain works with math that are very suitable for physics."

One of Grabowska's most memorable scientific experiences occurred as a graduate student, working with thesis advisor David B. Kaplan. Together, they had an idea about how to use lattices to study theories in which the mirror image of a phenomenon differs from the original image, called parity violation. Such theories oppose the intuitive idea that right and left should be identical in nature. But, in fact, our intuition is wrong. Right and left are not identical in some observed subatomic phenomena. So it is of fundamental importance to understand such theories. "We really got into it," Dorota recalls. "We thought we had a way to figure it out. There wasn't like a single 'aha' moment. But over the course of a couple of weeks, we were seeing it all come together. I would start going home, and David would start walking home, and I would have an idea, and I would call him up, and we'd talk on the phone about ideas as dusk was coming. The two of us were in this bubble of trying to understand how something works. And something that no one else had really thought about."

Grabowska and Kaplan still don't know whether their idea of using lattices to understand parity violation is right or wrong. Mathematically proving its correctness is extremely difficult, and simulating the proposal on a computer will have to await the further development of quantum computers. "It wasn't exciting because it was right," says Dorota. "I knew that even if we thought we were right, there were checks we couldn't do. But it was the coming up with something together and figuring out something together, the way that this could work. It was kind of an *uncovering.*"

Dorota wants to bring more diversity into the enterprise of science. In 2020, they won the Leona Woods Distinguished Postdoctoral Lectureship Award, designed to celebrate the scientific accomplishments of outstanding female physicists, especially those from underrepresented minority groups. In remarks on receiving the award, Grabowska said, "It is no secret that there is a severe underrepresentation of women and people of color in physics, as well as of people who are out and visible as members of the LGBTQ community. Such awards offer those from traditionally underrepresented and marginalized communities the opportunity to shine and to claim that we do belong." Indeed, only about one-sixth of all physicists are women. By contrast, the fraction of women in biology is almost 50 percent. It's harder to get reliable statistics on LGBTQ scientists.

In the fall of 2022, after three years at CERN, Dorota moved back to the University of Washington in Seattle, where they had done their graduate work. Dorota bought a condo there. It's a ten-minute walk to the bus stop, and from there a bus ride to the university. Dorota lives alone. "I don't mind living alone, but sometimes it would be nice to have a fuzzy old cat say hi. When it gets cold and dark outside, it would be nice to come home to something cute and fuzzy." But Dorota travels too much to keep a pet.

The apartment, one bedroom and one bath, is part of a 1928 Tudor-style building, with nine units all on the ground floor. Dorota has filled the place with art, prints, and knickknacks, including a carved dog from South Africa, paintings made by friends, and paintings given them by their parents. "It's busy," says Dorota, "but it's collections that remind me of people and experiences." Then there are the framed insects—beetles and butterflies. "I think insects are pretty," they say. "They're fascinating. They're so different. Something nature made. I see a grasshopper and I want to follow it."

"What do you do for fun?" When that question was posed to neuroscientist Lace Riggs, she replied that nothing was more fun than being in the lab.

"Physics is not the most fun thing I do," says Dorota. "Physics is sometimes very frustrating. Sometimes, it's just a slog, where you're just beating your head over something. The thing I do for fun is that I'm a rock climber. I'm usually found at the climbing gym or outdoors. Most of my non-physics time is dedicated to climbing. Most of my friends are also climbers."

When asked if they work on weekends, a good barometer for how consuming a profession is, Dorota replied that "as a grad student, I used to work on the weekends. And then, as I progressed up, I stopped being able to work on the weekends. That doesn't mean my brain turns off on the weekends. I might jot something down. If an inspiration hits, or if I really need to get something done, like a paper deadline, or a student really needs something, I will work on the weekends. In general, I don't work on weekends. I think it makes me a better scientist, because it gives my brain time to percolate. If I work on the weekend, the next week is hard. I want a life that feels balanced. During the climbing season, I'm off in the mountains for most of the weekend. I want time to see my friends. I want a life that isn't just doing physics."

At the University of Washington, Dorota, now thirty-seven years old (as of 2025), has the position of research assistant professor. It's not a tenure track position, meaning that it does not automatically lead to a permanent job at the university. Getting tenure is the holy grail for most young academics. "I'm not yet secure in physics," says Dorota. "Anyone who doesn't have tenure is worried about getting tenure. We have so many brilliant people doing PhDs and postdocs not getting faculty [tenure track] jobs. When I'm by myself with my notebook, figuring things out, then it's perfect. I'm confident. I trust that I will eventually get there. I trust my own instincts. But when it comes to impressing others, and impressing hiring committees . . . I realized just how much luck comes into getting tenure. It's what jobs open up, what the funding is like, whether you were in this meeting or that meeting, whether you had this or that faculty member pay attention to you."

Independent of their abilities in theoretical physics, Dorota has made some particular life choices that make the difficult road to tenure even more difficult. They struggle with those choices even while they embrace them. "There's always the possibility that if I worked on weekends, I might get tenure," they say. "But for me personally, I can't work on weekends. Part of it also for me is that I don't want necessarily to leave the area that I am currently living in. There are personal ties here, and I'm a rock climber. I need to be around rocks, at least during the sunny months. Also, I'm part of the LGBTQ community. There are not a lot of places that I would be comfortable moving to. Being a physicist is my job, but I also want to have a healthy, happy life. So, it's a question of how do you prioritize all that."

For many scientists, their scientific work is the top priority of their lives, even though they may have spouses and partners, children, hobbies. For others, like Dorota, alternative priorities equal or even exceed their work in science.

Both kinds of scientists can make significant scientific contributions. However, it is also true that when a scientist is in the middle of a research problem, whether in the laboratory or embedded within pages of mathematical calculations, the work is all-consuming. From our experience and that of the many scientists we've known, the outside world completely disappears when a scientist is enveloped within the throes of a particular research problem. The same is arguably true of painters, musicians, writers, and other creative people. During the creative moment itself, time and ego, body and place all vanish, leaving only the pure sensation of focus and being.

There is something else to learn from Dorota. Dorota Grabowska vividly illustrates the job insecurity that many young scientists experience, especially those in the academic world. There the university systems dictate something akin to a six- or seven-year probation period, and positions become permanent only after a strongly positive review of one's teaching and research. It takes perseverance and commitment to endure such uncertainty.

BARBARA McCLINTOCK

On May 13, 1947, Barbara McClintock rose early to finish the spring planting of her corn, an annual ritual for the past twenty years. "I was so interested in what I was doing I could hardly wait to get up in the morning and get at it," she once said. Now, from her lumpy acre of land, she could look in one direction and see her laboratory window at the Carnegie Institution's Department of Genetics. In the other direction, she could gaze out to the water and beyond to the little village of Cold Spring Harbor on Long Island. Most likely, she would have been wearing baggy pants and a short-sleeved white shirt. Forty-four years old, with short, tousled hair, wire-rimmed glasses, and an impish smile, she cut a slight figure, less than a hundred pounds altogether. Who could have guessed from her looks and demeanor that she was one of the greatest biologists in the world, the foremost authority on the genetics of maize, only the third woman to have been elected to the National Academy of Sciences, the first female president of the Genetics Society of America? A hint of such power might be glimpsed in her eyes, which radiated a piercing and fearless intelligence.

Her most important discovery, for which she would belatedly win a Nobel thirty-six years later, was in progress.

Alone, she walked the rows of her seedlings. Each of her several hundred plants she knew personally. She knew the parents she had mated to produce the plant, she knew the genetic makeup of the parents, she knew their chromosomes under the microscope. (Chromosomes are the long, thread-like molecules that contain the DNA and genes of the cell.) A wooden paddle stuck in the ground by each seed identified its heritage. Over the course of the growing season, from May to October, she would come every day to this field to water her plants, to nourish them, and to peer at their patterns of colors and stripes and waxiness. Indeed, she was famous among her colleagues for her extraordinary powers of observation. When the time came to mate the new generation, she would take all the precautions of the most meticulous matchmaker to ensure that the right pollen fertilized the right eggs. McClintock did all the fieldwork by herself, even the routine tasks. She didn't trust anyone with her plants, and she didn't tolerate fools. As she once wrote to a colleague, the Department of Genetics "had a greenhouse man, but he is not too bright."

From a young age, McClintock was both prickly and proud, fiercely independent. She was born Eleanor McClintock in June 1902, in Hartford, Connecticut, the daughter of Thomas Henry McClintock, a physician, and Sara Handy, a Boston Brahmin. According to McClintock's accounts, her parents decided that Eleanor was too delicate and feminine a name for her temperament and changed it to Barbara. As she said to

Evelyn Fox Keller, who conducted extensive interviews with her in the late 1970s, "My mother used to put a pillow on the floor and give me one toy and just leave me there. She said I didn't cry, didn't call for anything." McClintock grew up a loner, content with her solitude. As a child, she loved to sit by herself, "just thinking about things." Her uncle bought a truck, and McClintock said her early interest in motors (and the hidden structure of things) began by watching her uncle work on that truck. "My father tells me that at the age of five I asked for a set of tools. He did not get me the tools that you get for an adult; he got the tools that would fit in my hands, and I didn't think they were adequate. . . . I wanted *real* tools, not tools for children."

At Erasmus Hall High School, in Brooklyn, New York, McClintock realized that she liked science above all other subjects. "I would solve some of the problems in ways that weren't the answers the instructor expected. . . . It was a tremendous joy, the whole process of finding that answer, just pure joy."

During McClintock's freshman year at Cornell, she refused to join a sorority when she realized that some people were invited and some not. As she recalled, "Here was a dividing line that put you in one category or the other. And I couldn't take it." A year later, at the age of eighteen, she became a standout maverick on campus overnight by having her hair cut short. In her junior year, McClintock took a genetics course and loved it. In her senior year, she joined a jazz improvisation group and played tenor banjo.

McClintock stayed on at Cornell as a graduate student. By this time, she knew that genetics was her passion, her self-fulfillment, her life. She had friends, but she kept them at a distance. Years later, she told Keller, "There was not that strong necessity for a personal attachment to anybody. I just didn't feel it. And I could never understand marriage."

By 1934, the string of research positions at Cornell had dried up. McClintock, now a world-renowned geneticist, had no financial support and no job offers. She grew bitter, blaming much of her trouble on being a woman in a man's world. "At this stage," she recalled, "in the mid-thirties, a career for women did not receive very much approbation. You were stigmatizing yourself by being a spinster and a career woman, especially in science." In fact, the professional opportunities for women in science in America did not much improve until after World War II, and then only slowly.

Eventually, in 1942, McClintock received a research position at Cold Spring Harbor Laboratory on the North Shore of Long Island. This private institution had been founded in 1890 as a laboratory devoted to plant biology and biomedical research, as well as biology education, funded in part by the Carnegie Institute of Washington. At the time McClintock joined, the facility had a close connection to the Biology Department at Columbia University. Many leading geneticists flocked to Cold Spring Harbor during the summer, but only a handful remained during the rest of the year. McClintock was the only maize geneticist there.

As was common for cell geneticists but uncommon for most other biologists, McClintock performed both the breeding work in the field and also the laboratory work, studying the microscopic chromosomes that contained the genes. Before her work at Cornell in the late 1920s and early 1930s, the principal organism for such studies was *Drosophila,* the fruit fly, useful because of its rapid life cycle. Maize, on the other hand, had the advantages of highly visible genetic traits, such as kernel colors and leaf markings, and its chromosomes were larger than *Drosophila*'s and thus more easily studied under the microscope. (Eventually, large chromosomes in the salivary gland cell of *Drosophila* were found.) Refining new cell-staining methods, McClintock was the first person to identify and characterize the ten different chromosomes of maize. By the early 1930s, she had made maize, or *Zea mays,* of equal importance to geneticists as *Drosophila.*

Evidently, McClintock had exceptional powers in observing the innards of cells under the microscope. The geneticist Marcus Rhoades once told McClintock that he marveled at the way she could see so much when she looked at a cell under the microscope. Her reply was "Well, you know, when I look at a cell, I get down in that cell and look around." In a description very similar to Lace Riggs's comments about observing and "feeling" individual neurons (chapter 3), McClintock recalled that "the more I worked with [the chromosomes], the bigger and bigger they got, and when I was really working with them, I wasn't outside, I was down there. I was part of the system. It surprised me because I actually felt as if I were right down there and these were my friends."

Despite her extraordinary "friendship" with the tiny residents of cells, McClintock had periods of self-doubt. Even after being elected to the prestigious National Academy of Sciences in 1944, McClintock recalled of her research at that time, "I was really quite petrified that maybe I was taking on more than I could really do. . . . I got very discouraged and realized that there was something wrong. . . . I wasn't getting things right at all."

—

What McClintock was studying at that point in her career was the way in which maize chromosomes could break in particular places and then rejoin, in the process introducing mutations and changes in genetic traits, such as colors and stripes on the corn kernels. By careful breeding, she found that she could produce plants whose chromosomes *regularly* went through these breakage-fusion mutations.

A chromosome in a string of genes. Each gene is a molecular instruction, spelled with a long and particular sequence of four molecular "letters": adenine (A), cytosine (C), guanine (G), and thymine (T). The average gene contains about 2,800 of these letters. In a conventional mutation, the order of these letters changes, usually by copying error in reproduction, causing a malfunction in the action of the gene. Mutations occur in the standard theory of inheritance, but, before McClintock's work, they were thought to be permanent rather than transient, and they were also thought to be random.

Evidently, McClintock's breakage-fusion mutations were not random. Something was altering the genes on the maize chromosome in a regular and systematic way. That idea, already, was a revolution. Until then, biologists had viewed genes as fixed links on the chain of the chromosome, unchanging except for random mutations, sending information and instructions in a one-way path from the chromosome to the rest of the organism.

McClintock spent the next several years trying to find out what was controlling the orderly mutations in corn. During this period, she became obsessed with the problem. She worked day and night on it, almost always alone, sometimes sleeping the fitful nights on a cot in her lab.

Now, on the late spring morning of May 13, 1947, as she stood among her new plantings, McClintock felt that she was close to the answer of what controlled the regular mutations in corn. She collected her four-by-six-inch index cards, on which she would write the pedigrees of each plant, and returned to her lab. In a few months she would look at her new offsprings' chromosomes under her microscope. And there she would make the greatest discovery of her life. McClintock found that genes actually changed positions on the chromosomes, rearranging themselves in a controlled way. This was a new kind of mutation: not the changing of molecular letters within a gene, but a change in the position of the gene on the chromosome.

No longer could one think of genes as fixed in position, or of the chromosome as a static warehouse of instructions. The chromosome and the genes on it were a dynamic system, changing during a single lifetime, both controlling and being controlled by the rest of the organism. At the time, McClintock fathomed some of these ideas, but not all. Even today, biologists don't understand the details of how information from the developing organism is relayed back to the chromosomes, with the potential of altering the further evolution of the organism.

Barbara McClintock was emotionally involved with her plants. As she told Keller, "I start with the seedling, and I don't want to leave it. I don't feel I really know the story if I don't watch the plant all the way along. So I know every plant in the field, I know them intimately, and I find it a great pleasure to know them."

Contrary to the image of science as a sterile dispassionate endeavor, most scientists become emotionally involved with their work. In fact, as discussed in chapter 4, good science *requires* that involvement, in order to spend the days and nights and months and years in the laboratory or study. Good science requires the complete immersion within a project, to understand, to find answers, to satisfy an urgent desire to know.

Barbara McClintock exemplifies the scientist whose entire life seems to be built around her scientific career, almost to an obsession. Her ability to work alone and still make significant progress was, in large part, a reflection of her times. Today, the equipment used by biologists is far more advanced than McClintock's simple microscope, and most biologists work in large groups. Another distinctive feature of McClintock's era: Female scientists, even biologists, faced many more obstacles and more gender bias than female scientists today. For a female scientist to succeed in the 1920s through the 1960s, as did such people as Barbara McClintock and astronomer Vera Rubin and physicist Lise Meitner, a great deal of determination and independence was needed.

Chapter V

WHAT GETS THEM STARTED?

In the spring of 1817, at age eight, Charles Darwin (1809–1882) was sent to a school in Shrewsbury, a private boarding school for the local aristocracy and the bourgeoisie. His mother was to die only a few months later, probably from a severe ulcer or stomach cancer. In his autobiography, Darwin recalled that "my taste for natural history, and more especially for collecting, was well developed. I tried to make out the names of plants . . . and collected all sorts of things, shells, seals, franks, coins, and minerals. The passion for collecting which leads a man to be a systematic naturalist, a virtuoso, or a miser, was very strong in me, and was clearly innate, as none of my sisters and brothers ever had this taste. . . . Looking back as well as I can at my character during my school life, the only qualities which at this period promised well for the future, were, that I had strong and diversified tastes, much zeal for whatever interested me, and a keen pleasure in understanding any complex subject or thing. . . . Early in my school days a boy had a copy of the 'Wonders of the World,' which I often read, and disputed with other boys about the veracity of some of the statements; and I believe that this book first gave me a wish to travel in remote countries, which was ultimately fulfilled by the voyage of the 'Beagle.' "

The Nobel Prize–winning American chemist Linus Pauling (1901–1994) also exhibited his interest in science at an early age. In 1916, he often visited an abandoned smelter near Oswego, Oregon. The smelter was his playground. He found an old testing laboratory, a place iron ore had been assayed. The shelves were full of dusty ore samples, bottles of concentrated acids and other chemicals.

The young Pauling would bring a battered suitcase and pack up some of these intriguing chemicals to take home—pounds of potassium manganate, gallons of nitric acid—and he made a lab for himself in the basement of his mother's boardinghouse. Pauling was amazed by his own experiments and those of a friend. As Pauling later wrote, "I was simply entranced by chemical phenomena, by the reactions in which substances, often with strikingly different properties, appear; and I hoped to learn more and more about this aspect of the world."

Scientists, in fact, emerge from a wide variety of backgrounds and early experiences. Sometimes, their parents are scientists, sometimes not. Sometimes scientists-to-be come from affluent backgrounds, sometimes not. For Darwin and Pauling, their future careers were fairly evident in childhood. Although there are exceptions, most budding scientists express an early interest in exploring the physical world, or in numbers and the quantitative measure of things. Most are independent-minded from a young age. And most receive some kind of encouragement from a parent, a teacher, or a friend. A range of these beginnings and early motivations can be found in the lives of the various scientists in the interchapter profiles in this book, including physicists Dorota Grabowska and Werner

Heisenberg and John Mather, biologist Barbara McClintock, neuroscientist Marta Zlatić, ecologist Magdalena Lenda, and astronomer Govind Swarup.

German American scientist Joachim Frank, who won the 2017 Nobel Prize in Chemistry, grew up in Siegen, Germany. His father was a judge. His mother stayed at home, taking care of her four children. His elementary school was right across the street from his house. As he recalls, "I was eight years old when I started my first experiments in the dark place underneath the veranda where our little pig once roamed. It was natural curiosity that made me do it, before I had any concept of science." Like Pauling, the young Frank liked to gather various chemicals and experiment with them. "I built a shelf, collected little *Magenbitter* liqueur bottles and filled them with every liquid I could get hold of: oil, water, gasoline and, when I was a little older, hydrochloric acid. In bouts of intuition, I mixed the fluids, exposed metals to them and recorded the results. I watched calcium carbide dissolve in water and enjoyed watching the violent reaction and the smell of the escaping gas. I watched zinc dissolve and bubble up in hydrochloric acid. I heated up coal in a metal container connected to a tube since I'd heard that a flammable gas would escape."

Japanese physicist and electrical engineer Shuji Nakamura, who invented light-emitting diodes, was born in Oku, a tiny fishing village on the Pacific coast of Shikoku, the smallest of Japan's four main islands. Besides fishing, farming is the main occupation of the village, and Shuji's maternal grandparents were farmers. Shuji's father worked in maintenance for Shikoku Electric Power. From him, Shuji learned how to make wooden toys, such as catapults and bamboo propellers. As with many budding scientists, Shuji enjoyed building things. In his biography, Shuji says that he constantly fought with his bigger older brother and always lost. But "though physically defeated, mentally [I] would never give in." As a schoolboy, Shuji's passion was volleyball. His school had no gym, so he had to practice outside in the mud. In his biography, he says that he was "fiercely competitive from an early age and always hated to lose." This competitiveness and striving to win seems a common feature of many budding scientists. Shuji does not relish being slighted. At age thirty-four, with years of hands-on experience in the lab but without having published a single paper, Shuji spent some time at the University of Florida. The younger graduate students there treated him as a mere technician when they discovered that he didn't have a PhD and no publications under his belt. This condescending attitude, plus Shuji's innate competitiveness, only fueled his ambitions. "I feel resentful when people look down on me," he said. "At that time, I developed more fighting spirit—I would not allow myself to be beaten by such people."

Some scientists do not develop a particular interest in science in childhood but have other experiences that engender an independence of mind. Molecular biologist Carol Greider won the 2009 Nobel Prize in Physiology or Medicine for her discovery of the enzyme that helps prevent chromosomes from losing their ends in cell divisions. (The Nobel Prizes do not have a separate category for biology but include it under the heading of physiology or medicine.) Greider remembers walking to school in Davis, California, as giving her a sense of freedom. "This early responsibility was something that shaped my sense of independence. For me school was something that was a kid's responsibility. Parents were not really involved." The early death of her mother in 1967 further shaped her need for independence of mind. "This event played a major role in my learning to do things on my own." Greider and her brother and friends made up games like tapping out a code on the radiators and sending notes on string outside the kitchen windows to communicate. But she says that "unlike many scientists I know, I was not a kid who knew from early on that I wanted to be a scientist. I think one important thing I learned in my early years was to focus intently on the task at hand, such as learning German when we were in Heidelberg, to the exclusion of other things going on around me. This survival skill served me very well in later years. Focusing on certain goals and ignoring obstacles came naturally to me." Greider recalls that in high school, "I never considered myself one of the smart kids, they seemed confident and driven. I just enjoyed learning and especially spending time with friends." She did do well in biology and "was particularly captivated by my twelfth grade biology class, where we learned a lot of physiology from a very motivated science teacher who had a Ph.D. I loved learning new material and being challenged, so I decided to major in biology in college." Thus began her career in science, possibly ignited by an inspirational high-school teacher.

Many scientists show an interest and aptitude for mathematics at an early age. Physicist Don Page was such a person. He grew up in Alaska, where both of his parents were elementary-school teachers and fundamentalist Christians. The Bureau of Indian Affairs ran schools in small villages, and Page's parents would go out to these villages and end up providing many services in addition to teaching school, such as delivering mail and giving shots, since these remote villages were far from doctors. Page's parents were quite independent, a characteristic that they may have transmitted to their son. Both had a college education, and his mother was "fairly mathematical." As he recalls, "I can still remember a big board, maybe about two feet by three feet, on which my father had written all the numbers out from one up to 100. . . . It was somewhere between the ages of 3 and 6. . . . I still sort of visualize the numbers from one to 100 more or less how they're arranged on this chart. . . . In the beginning of high school, I got some of the *Advanced Mathematics Made Simple* books. Most of those things in math I probably learned from books before they were really covered in school." Page's research has explored the speculative areas of cosmology, in collaboration with, among others, Stephen Hawking, and he's retained his fundamentalist beliefs. There are many scientists—especially physical scientists—who are religious, adopting the view that science and religion are, in Stephen Jay Gould's parlance, "non-overlapping magisteria" that can peacefully coexist. However, most scientists of faith are not fundamentalists.

Some future scientists express an interest or aptitude at an extremely early age. Thomas Lozito, assistant professor of stem cell biology at the University of Southern California, got his start in science at age three, when he first laid eyes on a box turtle. "It was love at first sight," he says. "I was obsessed." His dad, a real estate agent, discovered the turtle in a house he was selling. Young Lozito began collecting other turtles, frogs, lizards, and snakes. To the dismay of his mother, he created a home for his reptiles in the family den, transforming it into a jungle. "My goal was to keep and breed every species." Today, in his lab at USC, Lozito studies how lizards and salamanders regrow missing limbs.

In our own cases, none of our parents were scientists, but we were inquisitive about the physical world. Alan Lightman's father was a businessman who managed movie theaters, and his mother a dancing teacher. He loved to build things, like rockets and remote-control devices. He created his own laboratory in a large closet, stocked with electrical equipment and chemistry glassware. He also had an aptitude for mathematics. Martin Rees's parents were both teachers, although not of science, and ran a small country village school. He didn't have any particular ambitions in science, although he was good at mathematics, and on seaside holidays he was fascinated by the tides: Why did high tide occur at different times along the coastline? He also puzzled over mundane kitchen phenomena, such as: When the water in a circular washing bowl is set spinning, why do the tea leaves form a little pile at the center?

Finally, some future scientists launch themselves into their careers not because of an early interest in science or mathematics, and not because of encouragement from their parents, but in reaction against adversity, somehow summoning up within themselves a desire to beat the bad odds of their childhood and even to address those difficult conditions as a professional. Such was the case with neuroscientist Lace Riggs (see chapter 3).

A not dissimilar motivation occurred in the life of Chinese chemist Tu Youyou, who won a Nobel Prize in 2015 for the discovery of a drug, artemisinin, used to treat malaria. Artemisinin has saved millions of lives worldwide. Tu was born in Ningbo on the east coast of China, where her father worked in a bank. Her mother looked after her four brothers and her, the only girl in the family. At the age of sixteen, Tu contracted tuberculosis and had to take a two-year break from school and receive treatment at home before she could resume her studies in high school. Tu recalls that "this experience led me to make a decision to choose medical research for my advanced education and career—if I could learn and have (medical) skills, I could not only keep myself healthy but also cure many other patients." Unlike many professional scientists, Tu never received any advanced degree and never studied abroad. In the middle of the Cultural Revolution, the China Academy of Traditional Chinese Medicine appointed her to head a research group to search for antimalarial drugs among traditional Chinese medicines. "As a young scientist, I was so overwhelmed and motivated by this trust and responsibility," she recalled in her Nobel address. "I also felt huge pressure from the high visibility, priority, challenges as well as the tight schedule of the task." In Tu's case, it was evidently a combination of her personal illness as a child and the expectations of her society that drove her in science. And very, very few scientists, either pure or applied, can claim that their discoveries have saved millions of lives.

MARTA ZLATIĆ

Some people have a tin ear for language. Others have a remarkable facility with language and can pick up a new one as easily as falling off a log. Neuroscientists have found that people with such language ability have stronger connections in the white matter of the brain, lying beneath the gray matter. Marta Zlatić is one of those super-connected people. In addition to her native Croatian, she is fluent in English, German, French, Russian, and Spanish. For icing on the cake, throw in ancient Greek and Latin. In fact, Zlatić is a neuroscientist herself. And it was her interest in language that got her there. "Towards the end of high school," she says, "I was quite torn between natural science and linguistics. Then I thought maybe I'll go into trying to understand how the brain generates complex behavior such as language. That was the motivation—to marry these two passions."

Her passion for biology began with her interest in animals. "I was always fascinated by nature and by animals, especially their behavior," she says. "I had this excellent biology teacher in middle school who was very inspiring. As a child I played outdoors a lot, and I liked observing animals—insects, birds, mammals, fish, mollusks." Marta did not have any conventional pets because her family traveled too much. But she did have a pet turtle, named Đina, who lived with the family for many years. "I liked watching Đina and trying to train it. It certainly learned. It became very tame. It learned to recognize me. It even learned to follow me around the room."

Marta grew up in an intellectual family. Besides her father, Professor Veljko Zlatić, a theoretical physicist at the Institute of Physics in Zagreb, her mother, Nena Franičević-Zlatić, studied philosophy and French and worked for the Ministry of Culture. "My parents were incredibly nurturing and inspired in me the love of learning, a curiosity about the world, and a fascination with both nature and arts and literature."

Since Marta was torn between linguistics and biology, she decided to pursue two BA degrees in parallel: natural sciences (specializing in neuroscience) at the University of Cambridge and general linguistics and Russian at the University of Zagreb. In between terms in Cambridge, Marta would hurry back to Croatia for her studies in linguistics. In her fourth year she finally had to give up linguistics because she had to spend more and more time in the UK.

While at Cambridge, Marta pursued yet a third passion, the theater. Combining her interest in ancient Greek with her interest in acting, she starred in several of Euripides's and Sophocles's tragedies, as well as in plays by Shakespeare, Racine, and others.

By 2000, Marta had made up her mind to pursue neuroscience and enrolled in a PhD program at the University of Cambridge. "Originally, I thought I would study the neural basis of higher brain functions, such as language. However, I changed my mind because I realized we didn't even understand how an insect flies, let alone how a human speaks, so I wanted to first study simpler and more tractable behaviors in smaller brains." Like the behavior of insects.

The brains of humans and other animals consist of brain cells, called neurons, with many connections between them. (See chapter 3 for a longer discussion of the brain.) For some years, neuroscientists have known that particular groups of neurons are responsible for particular kinds of movements, responses to sensory input, and other behavior. The connections between neurons—which neurons communicate with which other neurons—play a critical role for all of these brain functions.

During her PhD and postdoctoral research, Zlatić studied the development of sensory neurons in the embryos of *Drosophila,* commonly known as fruit flies. As a brain develops, the neurons create long filamentary extensions called axons and dendrites, which make highly targeted and specific connections with other neurons. Each neuron sends chemical signals to other neurons via its axon and receives signals from other neurons via its multitude of branch-like dendrites. Zlatić found out how particular axons corresponding to particular senses, such as sound and touch, make connections with specific target neurons. In addition, she discovered particular proteins that control how the sensory axons find their targets.

After a few years of postdoctoral research, Marta obtained her own independent position as a principal investigator at the prestigious Janelia Research Campus of the Howard Hughes Medical Institute in Loudoun County, Virginia, where she spent the next ten years, from 2009 to 2019. There she switched from studying how neural circuits develop to studying how information received by sensory neurons is integrated by the brain and used to learn about the world and make decisions. Her lab developed new approaches for bridging the gap between behavior, neural circuits, and genes. Together with the lab of her husband, Dr. Albert Cardona, they mapped the structure of the circuits that underlie learning and decision-making. New tools that genetically targeted neurons and activated them with light-sensitive chemicals allowed Zlatić to activate and deactivate one group of neurons at a time and identify key neurons that drive learning and decision-making. (Such techniques, called optogenetics, use proteins that can be activated by light. These special proteins also have the ability to bind to particular neurons.)

Many biologists eventually become leaders of laboratories and supervise the work of graduate students, postdoctoral fellows, and other researchers in their lab. Marta's journey began at Janelia. Marta is wistful when she recalls her ten years there. Because Janelia is endowed and has its own internal funding, no grant writing is required, a constant source of anxiety for most biologists. "It was a wonderful place to start a lab," she says. "Also, the whole idea was about synergy and collaboration."

At Janelia, Zlatić's first research group had two people, then six. As she recalls, "When you start running your lab, the [leadership] skills you learn by trial and error. The most important skill is communication with people. It's learning to let go. It's learning that it's now not only you doing the work. You do it with them. So, not to micromanage. Initially, it's your project, but then it becomes their project. You have to learn how to inspire without controlling." Evidently, the skills needed to lead a team in science are the same as those in businesses, universities, hospitals, and many hierarchical organizations.

Marta has been recounting her career when suddenly she stops mid-sentence. "Can I take one second? I have to check whether my son has switched off the lights." It's nine thirty p.m. Marta's children go to bed at nine p.m. each night, but her son, age eleven, likes to read for an extra half hour. It seems that her daughter, age five, is already asleep. Marta says that she makes sure to get home around five p.m. most days. "I want to be here when they get home from school. They are both such wonderful little people and it is so beautiful to spend time with them. If I don't do that, I really miss it." She and her husband are involved with the kids for the next several hours, until bedtime. Then Marta goes back to work. That's on weekdays. She doesn't work on the weekends. "Weekends are just with the children," she says. "The kids are very engaging and such a joy to be with. Also, between shopping and tidying up and taking the kids to their activities, it's just physically impossible to work." Then there is the cooking. "My husband and I both cook. It's Mediterranean cuisine, similar to Italian."

With kids, Marta doesn't have time to act in plays anymore. But she does make time to go to the theater. On occasion, she plays tennis. In the summer, the family goes swimming, in the winter skiing. In the very late evening, after she has finished working, she has a tiny bit of time left for reading. Fiction mostly, she says, and mostly the classics, such as Dostoyevsky and Bulgakov.

A few years ago, Zlatić and her husband moved back to Cambridge to take up program leader positions at England's Medical Research Council Laboratory of Molecular Biology. Zlatić is also a director of research in systems and circuits neuroscience at the University of Cambridge. In her most recent work, she and her husband and other collaborators employed an electron microscope, a device that uses a beam of electrons rather than light to illuminate extremely tiny structures, to make a complete map of the brain of the *Drosophila* larva—all 3,016 neurons and the 548,000 connections between them. In electrical terms, this map constitutes a complete wiring circuit of the fruit fly larval brain. Prior to this landmark achievement, neuroscientists had fully mapped only three much smaller brains, each with only several hundred neurons: the nematode *Caenorhabditis elegans,* the larva of the sea squirt *Ciona intestinalis,* and the marine annelid worm *Platynereis dumerilii.* The complete map of a brain's circuitry is called a connectome. In principle, a connectome has all the information about the brain's organization, how the brain receives information, how it learns, how it stores memories, how it directs particular behavior. It's like having all the drawings for the architecture of an entire city, all the buildings and shops and connecting walkways and even underground tunnels. Everything about the city is there.

Zlatić and her collaborators' complete map of this insect city is a permanent resource that will be the basis for many theoretical analyses and follow-up studies in the future.

We ask Marta whether scientists will ever have a connectome of a human brain. It seems a long way to get from the fruit fly larva to the human brain, with some 100 billion neurons and 100 *trillion* synaptic connections. "I definitely think we will," she says. "The question is when. Right now, it is out of reach. But the methods for generating connectomes are increasing exponentially. The amount you can image is doubling every few years. And then, with machine learning, tracing neurons and their connections in such images is becoming automated. We will soon be able to compare many connectomes of closely related insect species that have different behaviors, to understand how behavior evolves. We will also be able to compare connectomes of many individuals of the same species to understand what determines individual differences between behavior—different personalities. We will soon also be able to do small vertebrate connectomes. Small fish, frogs, lizards. Eventually, we'll be able to do a mouse, and then a human."

Although scientists don't fully understand how memory is encoded in the brain, we do know that it must reside in the neurons and connections between them. And the connectome somehow contains all of that information. Marta goes on to say that with the connectome, "you might be able to read out the entire life experience of an individual and everything that individual has learned." This possibility sounds like sci-fi, both exciting and unsettling.

When asked whether, if you can "read out" the entire life experiences of a person, you could predict their response to a particular stimulus (like whether they would fall in love with a particular person), Marta says, "Responses to stimuli will always be probabilistic. You could predict responses with a certain likelihood. . . . I think there is inherently something probabilistic in behavior, because you have many neurons, and each one can fail." Could one say there is a 35 percent probability Robert would fall in love with Jane, and a 75 percent probability he would fall in love with Alice? Marta laughs, but she doesn't disagree.

What does she worry about? "I guess one always worries about one's kids," she says. "You can't be a mother and not worry. I remember when my son was born, I said, 'Okay, now life without worry is over.' Until then, I had no worries in my life, and I never thought about the future. It was only the joys of the present. That changes when you have a kid. I do worry about climate change and wars and what the world will be like for my children. I find it incredibly sad, that while we have so many good options of organizing the world in a much better way, for utilizing the resources in a way that does not destroy the environment and redistributing them in a fairer way, this is not happening. However, I do believe that humans are above all a social and cooperative species, and the key evolutionary advantages of our species, in addition to intelligence, are cooperation, language, communication, accumulation of knowledge, and empathy. I do believe that humans will end up surviving. I guess I am an optimist."

JOHN MATHER

John Mather recalls that his interest in science and mathematics began at an early age. At six years old, he realized that he could fill an entire page with numbers and never come to the largest possible number. "I saw what was meant by infinity," he remembers. His father, Robert, a scientist specializing in dairy cattle genetics, and his mother, Martha, a high-school French teacher, read him biographies of Darwin and Galileo. Because the family lived on a farm in rural New Jersey, in the rolling foothills of the Appalachians, books were hard to come by. Young Mather borrowed as many as he could from a bookmobile, which traveled around the county and visited the farm every couple of weeks.

"In high school, I poked around with a lot of cute things," Mather says. The boy loved to build gadgets, especially gadgets involving electronics. "One of the things I really enjoyed was building systems in the backyard to look at the sky. It was more fun to build them than to use them." There were no stores nearby, and the budding scientist could get parts for his inventions only by ordering them through the mail. He built a telescope from the Edmund Scientific catalog. Because of the Earth's rotation, stars in the sky slide off the field of view of a telescope unless it is outfitted with a device called a clock drive, which slowly and precisely rotates the telescope to compensate for the Earth's rotation. Young Mather built his own clock drive from scratch, using relays and capacitors and resistors and a little motor. "I was very proud of myself for coming up with this in high school," he says. "It was a pain in the neck to use it, but it was fun to build it."

Mather perfectly exemplifies the early beginnings of an experimental scientist-to-be. As a child he had great fun building things. In Mather's case, it seems pretty clear that his talent for invention was a native ability, as it is for many scientists. One is reminded of Mozart, who began composing music at the age of five; by age six, the child had performed before two imperial courts. But even future scientists of less talent than John Mather begin building and tinkering at an early age.

Young Mather also enjoyed competition. And triumph. He says that on various statewide tests, he was number seven in math and number one in physics for all of New Jersey. "That was encouraging," he recalls. With a borrowed microscope, he tried his hand at biology. But that subject didn't take. "Biology as taught in the ninth grade [ca 1961] was a lot of memorization. There wasn't much at the conceptual level, and it wasn't all that promising at the time. Biology looked too hard to me. But physics, I thought that's easier. You can do that by understanding. You don't have to remember everything."

By the time he went to Swarthmore for college, Mather was sure that physics was it for him. "I tried without much success to learn a little of the humanities and the arts, but even passing the courses in art history and music history was a challenge. In those courses I understood what other folks felt when they saw me doing so well in physics."

In 1970, while hunting around for a PhD thesis project, Mather met Michael Werner, then a new postdoctoral fellow working with Charles Townes (winner of the 1964 Nobel Prize, for masers), and Paul Richards, a young scientist with expertise in low-temperature physics. They were starting up projects to measure the radio waves left over from the Big Bang, the so-called cosmic microwave background radiation. "I liked all three of them immediately, as well as the proposed experiment," recalls Mather, "and I started right in. It was a new world for me, much more tangible than years of books and classes and late nights in the library."

The Big Bang model is much more than the hypothesis that the universe was created in a state of extremely high density and temperature some 14 billion years ago and has been expanding ever since. The model includes a detailed set of equations that precisely specify the density and temperature of the subatomic particles and energy of the universe at each moment in time after $t = 0$. One of the predictions of the model is that outer space should be filled with a smooth bath of radio waves, created when the universe was much younger and hotter. Further, the model specifies the amount of energy at each frequency of those radio waves, like the volume of each channel of a radio. In 1965, the predicted cosmic radio waves were indeed detected, but only in one channel. That was promising but not conclusive. To confirm the Big Bang model, the cosmic background radiation would have to be measured at many other channels, to see if the volumes (intensities) agreed with the predictions.

That's the project that Richards and his group were working

on. Their instrument performed fairly well, but its accuracy was diminished by the atmosphere of the Earth, which interferes with radio waves coming from outer space. Then they put a measuring instrument in a high-altitude balloon. That project didn't succeed. As Mather recalls, "This was the beginning of a baptism by fire, in the art of building instruments that would work in remote and hostile locations. It was a time to learn something of almost every area of engineering, from mechanical to optical to cryogenics to electronics."

Partly discouraged, Mather finished up his thesis, such as it was, and joined Pat Thaddeus's group at the Goddard Institute for Space Studies, in New York. "I was hoping to go into a new field of study. I thought that my work on the background radiation was awfully difficult, and it was going to be hard to do much better with balloons. I suppose I was reacting too much to the failure of the instrument. . . . Years later I threw many boxes of IBM cards into the trash, finally admitting defeat."

But then, in the summer of 1974, NASA announced its plan to launch satellite missions, and Mather suggested to Thaddeus that his failed thesis experiment would have worked better in space, aboard an orbiting satellite. Together with several other colleagues, Mather sent in a proposal to NASA. He was in his late twenties at the time. The proposal was funded, and that was the beginning of COBE, the Cosmic Background Explorer. It took fifteen years of work by many people before the satellite was launched, in 1989. But the work paid off. COBE measured the cosmic background radiation to unprecedented detail and accuracy in many frequencies, confirming the Big Bang model. The detectors that Mather and his colleagues built were so sensitive that they could measure the energy in different frequencies of the cosmic radio waves to one part in a hundred thousand, where tiny fluctuations in the waves hold keys to the tiny density of inhomogeneities in the early universe that later grew into galaxies.

"It's part of the great story we tell about where we came from. A lot of people want to know who their ancestors were. This is sort of an extension of that back as far as we can go, to the beginning of time. It also helps us think about our place in the universe—which has had a lot of importance in our philosophy for thousands of years."

In 2006 John Mather won the Nobel Prize in Physics for his work with COBE. One person who attended the prize ceremonies in Stockholm described Mather this way: "Tall, thin and bespectacled, Mather was a classic nerd. A team player to a fault, he could be righteous, even a little priggish, though he had a winning smile and lively sense of humor."

Mather shared the Nobel with another physicist named George Smoot, with whom he had collaborated on COBE. Until that evening in Stockholm, Mather and Smoot had been feuding for fourteen years. The reason: in 1992, Smoot prematurely announced their results, in violation of a team agreement. According to an article in *Time* magazine, "Mather was infuriated . . . in what [he] alleged was a grab for solo glory [by Smoot]." All a reminder that scientists, as much as everyone else, have egos, ambitions, and jealousies.

Mather has indeed been a team player, as are many experimental scientists working on large and complex projects, like Marta Zlatić, discussed in the previous profile. But Mather's team was much larger. COBE cost $140 million plus two thousand human years of labor. Such enormous collaborations require that the leaders have particular people skills and temperaments, different from those of loners like Barbara McClintock or Albert Einstein. As one of the team leaders for COBE, Mather had to work with a large group of scientists, engineers, and managers. "I wasn't born to that job," Mather says. "I got into it because I was the person who waved my hands and said, 'I've got an idea for a satellite mission.'" Referring to a book by Charlie Pellerin about how NASA builds teams, Mather says that scientific teams involve several quite different personality types. There are the "visionaries," like himself, the people organizers who hire and fire, and "the people who track down every detail." It takes all of these categories of people to make a working project. Mather's philosophy for leading teams is to let the scientists, like himself, do the science, let the engineers engineer, and the managers manage. "If you try to make a person do something that is way outside their comfort zone, it's not going to work well." Mather says he was always in and out of the offices of these other members of the team, discussing some new problem that had come up.

That intensity continued throughout COBE and afterward, when Mather was made the senior project scientist for another massively collaborative mission, the James Webb Space Telescope, launched at the end of 2021. "If you are riding the back of a tiger," he says, "you better not get off. There were plenty of times to worry about something. When you know that an entire team, and all of NASA, is depending on you, it does kind of get your attention. It's scary, actually. Now, I tell people, 'Never trust the boss, especially if you are the boss.'"

Mather's team spirit, and his generosity, can be seen in the words he wrote in the first pages of a popular book he coauthored with John Boslough titled *The Very First Light* (1996).

> I extend special thanks to my mentors over the years. . . . Paul Richards at Berkeley taught me experimental physics and designed the apparatus that started us on the road to measuring the cosmic background radiation spectrum. . . . Patrick Thaddeus at the Goddard Institute for Space Studies encouraged me to develop the COBE idea and pointed me toward the right people. Mike Hauser hired me at Goddard, watched over my entire professional career, and set me some nearly impossible tasks, all the while showing scrupulous attention to detail, compassion, and examples of gentlemanly behavior. . . . My engineering colleagues who managed and built COBE, the secretaries and business staff who made the NASA system work for us, my fellow scientists who defined the mission, guided the engineering effort, and analyzed the data, my friends at Headquarters who defended and funded the COBE—all were essential parts of the team.

When John Mather is not doing physics and astronomy, he likes to read, especially history. "Not so much the 'battles' kind of history," he says, "but how did we invent things over time. Every city should have a book for tourists that explains how this place was built. I don't want to know how they fought. I want to know how did they build that amazing cathedral." With his wife, Jane, who was a ballet teacher, John frequently went to see the ballet. And they loved to travel around the world. "I travel so that I can have a sense of what people are like when they are not here in the United States." Jane, twenty years older than John, died in 2022.

Now Mather is gaining a new family. In November 2023,

at the age of seventy-seven, he married again, to his "lovely neighbor," Cheryl. Childless before, he has now inherited two stepchildren, two stepchildren's spouses, and two grandchildren. "I am just really, really enjoying them," he says. "They're marvelous, brilliant, lively people."

Chapter VI

WHAT KEEPS THEM GOING?

In 1981, several years before his death, the great theoretical physicist Richard Feynman did an interview for the BBC television program *Horizon* in which he was asked about his Nobel Prize. Feynman's reply: "I don't see that it makes any point that someone in the Swedish Academy decides that this work is noble enough to receive a prize—I've already got the prize. The prize is the pleasure of finding the thing out, the kick in the discovery, the observation that other people use [my work]—those are the real things."

Scientists are motivated by many factors: simple curiosity, ego and personal ambition, the adrenaline of competition, the desire to understand the world, the impulse to create. And, as Feynman says, "the pleasure of finding the thing out." These motivations, in some form, drive all of us—lawyers, business entrepreneurs, teachers, medical professionals, architects. Is there any particular motivation unique to scientists? Before attempting to answer that question, let's hear from a few scientists on what keeps them going.

In an interview, Carolyn Bertozzi, winner of the 2022 Nobel Prize in Chemistry, describes her continuing fascination with science and its mysteries: "It's not hard to stay interested and curious in organic chemistry, especially in the way that it intersects with biology. Because there is so much we don't know about the natural world, including our own human bodies that we occupy every day. There's so much we don't understand about how our bodies work. We walk around every day in one of the great mysteries of life."

Moments of discovery, although few and far between, can indeed be thrilling and sometimes provide the motivation to keep at it for years seeking another such thrill, as when a golfer makes one great shot on the third hole and continues for another fifteen, hoping for another one. Here are James Watson's recollections of the moments when he realized that the structure of DNA must be a double helix: "As the clock went past midnight, I was becoming more and more pleased. There had been far too many days when Francis [Crick] and I worried that the DNA structure might turn out to be superficially very dull, suggesting nothing about either its replication or its function in controlling cell biochemistry. But now, to my delight and amazement, the answer was turning out to be profoundly interesting."

Molecular biochemist Carol Greider—winner of the 2009 Nobel Prize for her discovery of the sections of DNA called telomeres, which play a central role in the aging of cells—described the puzzle-solving aspect of science in her Nobel lecture: "The story of telomerase discovery is a story of the thrill of putting pieces of a puzzle together to find something new. This story represents a paradigm for curiosity-driven research and, like many other stories of fundamental discovery, shows that important clinical insights can come from unlikely places. [Here] I describe the process of scientific discovery—at times frustrating, at times misleading and perplexing, but yet also at times wonderfully exciting. The willingness to keep an open mind, to enter uncharted waters and try something new, along with patience and determination, came together to tell us something new about biology. Fundamentally this story shows how curiosity and an interest in solving interesting problems can lead to a lifetime of exciting discoveries."

Astronomer Sara Seager, a pioneer in studying the atmospheres of planets beyond our solar system, recalls a recent discovery when she and a colleague realized how to eliminate the problem of variability in the light of the parent star. (The chemical composition of a planet's atmosphere is revealed by how it absorbs the light of its central star as the planet passes in front of the star.) "We have all the telescopes at our disposal, but we're blocked from studying the atmospheres right now because of the stars. The stars themselves are variable. We had an 'aha' moment, my colleagues and me, when we were working on this stellar activity problem. As you move into the infrared [the long wavelength part of a star's light], the star's contamination shrinks dramatically. When we had this realization, we were in the midst of getting proposals done. I just feel so happy. It is the deepest satisfaction you can imagine. I don't know whether you've had a Zen moment with food. One time I had this perfect chocolate butterscotch mousse, with salty and sweet particles suspended in it. It was like taking a bite of that."

Chemist Danna Freedman describes such a moment: "Beginning in 2017, we decided to create molecules featuring the same information as defects in semiconductors. This was a tremendous challenge, and we weren't confident our approach would succeed. We tried molecule after molecule and for different reasons each of the molecules failed, yet we didn't lose hope; there were so many molecules and we learned something from each new failure. My student was sending me data constantly, but none of the data demonstrated the key property [we were looking for]. Then he sent me the absolute ugliest plot that I have ever seen in my life, but that plot showed the first sign of a signal. I responded by sending him a GIF of the happiest, bouncing llama I've ever seen and that's exactly how I felt. I felt like an exuberant llama with a giant open field in front of me."

Most often, science proceeds by small incremental steps, rather than by sudden giant leaps forward. The findings along the way can also keep a scientist going. Astronomer Henrietta Leavitt (1868–1921) worked for years at the Harvard College Observatory and was given the painstaking task of identifying and measuring stars whose brightness varies in time in a regular, periodic way. Over her lifetime, Leavitt identified and measured more than 2,400 variable stars. Even this apparently numbing and routine work, Leavitt found engaging. After a family crisis called Leavitt away from the observatory for two years, she wrote to the director, "I am more sorry than I can tell you that the work I undertook with such delight, and carried to a certain point with such keen pleasure, should be left uncompleted." Eventually, Leavitt found an important relationship between the intrinsic luminosity of such stars and their periods, allowing later astronomers to use them as distant indicators to measure the size of the Milky Way and to determine that the universe is expanding.

There's also the sheer fun of building new things. Rai Weiss, one of the world's great experimental physicists and winner of the 2017 Nobel Prize for detecting gravitational waves, spent more than forty years slowly improving the technology and sensitivity of his apparatus. When asked what kept him going for so long, he replied,

> I kept getting my charge every time we went back to the lab, things getting better and better. So I would spend weekends in the lab, and at night with the students, and we would make things work better. So it turns out that the thing which saved us all is that the experimenters got pleasure out of what they were doing. Turns out when you build things, it takes time to build them. Then you have to check them. And for me, it was a great pleasure to build them. I love to make electronics. You design the electronics and it doesn't work right away. Yeah, you've built it. You made it, and it doesn't do the thing it was supposed to do. Well, that's fun. There's a challenge right there. So you sit there and stew about it, and all of a sudden, yeah, of course if I'd done it the other way around this would've worked. And you try it and it works. And then you go out and have a beer with the guys.

Some scientists find pleasure in the sheer craftsmanship involved with building instruments. The late physicist Freeman Dyson made this observation about some students building a cryostat: "I happened to walk into a basement workshop at the bottom of the physics building at Cornell University. There I saw two students, dressed in the customary style, with bare feet and long, unkempt hair, building a cryostat for low-temperature experiments with liquid helium. . . . This was a new type of cryostat, working with the rare isotope of helium, that would take you down to a few millidegrees above absolute zero. The students were exploring a new world and a new technology. . . . The students were absorbed in putting this intricate maze of tubes and wires together. Their brains and hands were stretched to the limit. . . . At the time . . . they were not dreaming of Nobel Prizes. They were driven by the same passion that drove my grandfather, the joy of a skilled craftsman in a job well done. Science gave them their chance to build things that opened new horizons, just as their ancestors built ships to explore new continents."

And there is the simple pleasure of understanding. Of course, most of us, in all walks of life, experience the pleasure of learning and understanding. If that pleasure is any deeper for the scientist, it is because she or he may be the first person on the planet to understand some new property of the universe. Here Nobel Prize–winning chemist Roald Hoffmann writes about that pleasure: "The joy of understanding is one of the great emotional experiences of our life. In saying this, we do not belittle other pleasures—that of a child holding up a flower to a grandmother's face. . . . But the mind opening up—now that is a rare joy!"

Biologist Robert Weinberg describes his long-term satisfaction in being a scientist, especially when seeing other members of his research group make discoveries. "As a working researcher, many of the most gratifying moments occur when one's scientific trainees suddenly show deep insights that would never have occurred to them in their earlier years. Helping in the intellectual maturation of young people is surely as gratifying as many of the other facets of a scientist's life. The one eureka moment I have had in my own career came in 1978 when I realized that those in my lab had the ability to answer an age-old problem in cancer research: Can one demonstrate experimentally that cancer cells carry mutant genes that govern their aberrant behavior? This was most exciting, because proving this idea, which at the time was only a speculation, was within our reach experimentally!

"This euphoria lasted for several weeks until I became aware that someone had reported such work in a major research journal several months previously. Even without the eureka moments, the day-to-day work of a research scientist is endlessly stimulating and often fascinating. Solving simple and complex problems offers indescribable satisfaction, and I am most grateful for a career in which I was paid to have fun!"

Then there is the excitement of competition and rivalry that drives some scientists. Such forces are evident in this passage from Watson's book *The Double Helix:*

> By the time the hour-and-a-half train journey was over, Francis saw no reason why we should not know the answer [to the structure of DNA] soon. Perhaps a week of solid fiddling with the molecular models would be necessary to make us absolutely sure we had the right answer. Then it would be obvious to the world that [Linus] Pauling was not the only one capable of true insight into how biological molecules were constructed. Linus' capture of the α-helix was most embarrassing for the Cambridge group. About a year before that triumph, Bragg, Kendrew and Perutz had published a systematic paper on the conformation of the polypeptide chain, an attack that missed the point. Bragg in fact was still bothered by the fiasco. It hurt his pride at a tender point. There had been previous encounters with Pauling, stretching over a twenty-five-

year interval. All too often Linus had got there first. Even Francis [Crick] was somewhat humiliated by the event.

Undoubtedly, ego and ambition are motivations for many scientists, just as for nonscientists. In his autobiography, Nobel Prize–winning physicist Luis Alvarez (1911–1988) says that he had the longest Nobel Prize citation ever, was the first physicist to win the National Medal of Science, the first to use cosmic rays for a practical purpose, and the first civilian to land an airplane under radar control (using the first ground-controlled approach system, which he invented).

One could mention many more scientists who are driven by their personal egos. If they had gone into selling cars or landscaping or writing novels, they would have been impelled by the same forces. But it is important to make the point that many scientists are quite modest—for instance Kip Thorne, another Nobel Prize winner, who always insisted that the names of his graduate students went first on any papers he coauthored with them. Although Thorne was close colleagues with some of the greatest physicists in the world (including Stephen Hawking, to whom he offered sustained practical help), the walls of his office had no photographs of his posing with such eminences, no award certificates or commendations. At age thirty, he became one of the youngest full professors at the revered California Institute of Technology (Caltech), yet he never mentioned that fact.

Biologist Shinya Yamanaka gained international recognition in mid-2006 when he reprogrammed adult mouse cells to behave like embryonic stem cells without the use of embryos. Then he did the same with adult human cells. In 2012, he was awarded a Nobel Prize for this work. Back in 2008, after the important work but long before the prize, he said in an interview, "We were extremely lucky. I know many other scientists who have been working harder and who are smarter than we are."

Let's return to the question of whether scientists are driven by any motivation unique to their profession. Almost everything we see around us is impermanent. Forests burn down. Buildings crumble. Our bodies grow wrinkled and weak. Even the stars eventually use up their nuclear fuel and burn out. Faced with our mortality, consciously or unconsciously all of us want to leave something behind after our brief lives. For many of us, that legacy may be found in our children and grandchildren. For some of us, it may be a business we have built. For teachers, it may be their students. However, while all of these legacies live beyond our own lives, they too pass away. Our grandchildren may remember us, and our great-grandchildren may see photos of us, but in another few generations there will be not the slightest trace of most of us, or if there is a trace it will be one grain of sand on the beach of the internet. Few businesses last beyond a century. Our students may remember us, but that will be all.

But scientists have the possibility of leaving something long-lasting. In discovering new understanding about the physical world, scientists have the privilege of engraving a few lines in the platinum book of knowledge about the cosmos. Their names may not be attached, but they can have the deep satisfaction of knowing that they added something to human knowledge—not the kind of knowledge that shifts and tilts with new fashion or fad, not the kind of knowledge that varies from one culture to the next. But a universal knowledge, an eternal truth. The laws of quantum physics and relativity, the principle of natural selection, the structure of DNA are forever. It is almost certain that intelligent life across the far reaches of the cosmos would be familiar with these facts about the universe. Some of these eternal truths are big. Some are small. All bring immense satisfaction to their human discoverers. Here is physicist Melissa Franklin, describing her pleasure at helping to discover the mass of a subatomic particle called the top quark: "I like the act of measurement. I feel compelled to measure. It's more satisfying to find the top quark and measure its mass, which is going to be there forever. And it's going to be in a little book, or I guess they won't have books anymore, but it's going to be somewhere, forever, that measurement. Thinking back, that's very satisfying."

WERNER HEISENBERG

In his *Memoirs* (2001), Edward Teller describes a moment during his apprenticeship with Werner Heisenberg in the late 1920s. One evening Teller went to dinner at Heisenberg's bachelor apartment, where he was delighted to see "an excellent grand piano." A musician himself, and possibly trying to impress his mentor, Teller mentioned that he had been playing Beethoven and Mozart but was particularly fond of Bach's Prelude in E-flat Minor. At which point, Heisenberg sat down and performed the piece beautifully, even substituting a two-handed mezzo forte for the usual one-handed forte.

Music was Heisenberg's lifelong passion. So were philosophy and physics. All three passions are evident in an incident in 1920 when Heisenberg was nineteen years old, recorded in his memoir *Physics and Beyond* (1971). He had just finished playing some music with a friend when the friend's mother suggested that Werner become a professional musician. To which Werner replied, after bemoaning the decline of music, "In science, and particularly in physics, things are quite different. Here the pursuit of clear objectives along fixed paths . . . has quite automatically thrown up problems that challenge the whole philosophical basis of science, the structure of space and time, and even the validity of causal laws. Here we are on *terra incognita,* and it will probably take several generations of physicists to discover the final answers. And I frankly confess that I am highly tempted to play some part in all this."

Indeed he did. Only a few years later, at the age of twenty-three, young Heisenberg published his landmark paper on quantum mechanics, laying the foundation for that brand-new field of physics. In Heisenberg's mathematical version of quantum physics, each particle is represented by an array of numbers, called a matrix. The following year, the Austrian physicist Erwin Schrödinger proposed an alternative formulation of quantum mechanics, representing objects by continuous waves of probability rather than by matrices. Heisenberg's and Schrödinger's different formulations were shown to be equivalent, and both men won a Nobel Prize for the development of quantum mechanics.

Quantum physics, at work in our computers and smartphones and all modern electronic devices, has drawn back the veil shrouding the subatomic world. Quantum physics has presented a new picture of nature, and reality. Contrary to all our everyday experience, small particles behave as if they existed at several different places at the same time. Their positions and movements can be described only with probabilities, not certainties. This strange concept is part of the well-known Heisenberg uncertainty principle.

Heisenberg grew up in a privileged and highly educated family. His mother was a poet and the daughter of the headmaster of the Max-Gymnasium. The children of such families in Germany were always taught a musical instrument and learned the great classical works. Werner and his brother Erwin practiced daily, Werner on the cello, Erwin on the violin, accompanied by their father, August, who had an excellent operatic voice. Later, Werner took up the piano. One of his biographers, David Cassidy, plausibly conjectures that long hours of solitary piano practice, immersed in a world of order and harmony, must have contributed to the shaping of Werner's inner world.

Young Heisenberg also exhibited a strong will, and a desire to excel at all things. Although physically frail, he pushed his body as hard as he could. That began with athletics, an important part of German youth. Although the boy did not seem to be a natural athlete, he trained himself to ski over difficult terrain, to further challenge himself. A fellow student recalls Werner's running laps alone at the school with a stopwatch in hand to improve his time. Some years later, after being beaten by his students at ping-pong, Heisenberg proceeded to train intensively on a long boat trip from Shanghai to Europe. On his return, he could not be defeated.

Extremely protected by his parents, the young Heisenberg would stop speaking to anyone he thought had treated him unfairly, and he refused ever again to look at a certain schoolteacher who once slapped his hands with a rod. Like the young Einstein, Heisenberg developed an inner freedom, an independence, and a propensity to question all things. More than Einstein, he was a driven human being.

After attending the Maximilian school in Munich, Heisenberg went to the University of Munich to study with Arnold Sommerfeld, one of the great German scientists. He had already decided that he wanted to be a theoretical physicist. Most physicists are exclusively experimentalists, good with their hands and the building of things, or exclusively theorists, good with pencil and paper and mathematics. From a young age, Werner had realized that building things was not for him. As he wrote in his memoir, "I had never been really at home in the world of instruments, and the care needed in making relatively unimportant measurements had struck me as being sheer drudgery." Instead, young Heisenberg, philosophically inclined, wanted to solve the deepest questions about the nature of reality. Einstein had already shown that time and space are not what we think they are. There were other mysteries, in the realm of the atom.

Physics is the most philosophical of all scientific disciplines—concerned with the ultimate constituents of matter and energy, the fundamental nature of time and space, and the origins of the universe—and Werner Heisenberg was one of the most philosophical of physicists. He wrote an entire book titled *Physics and Philosophy* (1958). As a schoolboy he read Plato in the original and was particularly fascinated by Plato's *Timaeus,* in which the Greek philosopher distinguished between the physical world and the eternal world and conjectured that the ultimate constituents of matter are geometric forms, like cubes and tetrahedrons, rather than material objects.

Heisenberg recalled a conversation with his friends in the spring of 1920, shortly before he began college:

> Robert's references to Malebranche [a French philosopher] had convinced me that our experience of atoms can only be indirect: *atoms are not things.* This was probably what Plato had tried to say in his *Timaeus,* and, seen in this light, his speculations about regular bodies were beginning to make more sense to me. When modern sci-

> entists speak about the form of atoms, they must be using the word "form" in its widest sense, i.e., they must be referring to the atom's structure in time and space, to the symmetrical properties of its forces. . . . In all probability, such structures will forever elude our powers of graphic description, if only because they are not an obvious part of the objective world of things. But perhaps they are nonetheless open to mathematical treatment.

And, indeed, Heisenberg's work on quantum physics is extremely mathematical and abstract. His philosophical view was that particles have an existence only when we measure them. If a particle travels from A to B, and we measure its location at A and then B, we can say only that the particle existed at those two points. Between A and B, the particle does not exist in any meaningful way. Such an interpretation of quantum physics is as much philosophy as science.

By 1923, at the astonishingly young age of twenty-one, Heisenberg had completed his PhD in theoretical physics. His thesis advisor, Max Born, recalled that his student "looked like a simple farm boy, with short, light hair, clear bright eyes, and a radiant expression on his face. . . . His unbelievable quickness and acuteness of apprehension enabled him to do a colossal amount of work without much effort." A few years later, Heisenberg had his breakthrough in formulating the new quantum physics.

In 1927, at the University of Leipzig, Heisenberg began creating an international "school" of physicists, in a similar fashion to what Ernest Rutherford had done in Cambridge and Niels Bohr in Copenhagen. Edward Teller, whose recollection of Heisenberg's piano virtuosity was described earlier, recalls about twenty young men in Heisenberg's group, including himself (a Hungarian), some Germans, a few Americans, two Japanese, an Italian, an Austrian, a Swiss, and a Russian. Heisenberg, their leader, was twenty-six years old. Only a few years later, he was to be honored with a Nobel.

Some scientists, like Einstein, Freeman Dyson, and John Mather do little or no teaching and have no students. Others, like Heisenberg, build informal "schools," in which young scientists learn and work in an apprentice-like environment under the tutelage of an older master, similar to the art ateliers of past centuries.

In 1973, Heisenberg gave a lecture to the physics department at Caltech. The auditorium was packed. Several hundred faculty and students fidgeted anxiously in their chairs, everyone enthralled to meet a living legend in their field. The speaker was introduced. And then Heisenberg shuffled to the lectern. The person once described as looking like "a simple farm boy, with clear bright eyes, and a radiant expression" seemed like a weary old man, with a wrinkled face and a dark weight on his back. But that was not the biggest surprise. At the following reception for Heisenberg at the elegant Caltech faculty club, Richard Feynman, a professor at Caltech and a Nobel Prize winner himself, stood up and verbally attacked Heisenberg for the foolishness of his lecture, indeed mocked him to his face. In fact, Heisenberg's proposed "theory of everything" presented at the lecture did seem a bit nutty and grandiose, even to the graduate students in the audience. But there was something else. Underneath Feynman's scathing remarks, one could detect not only a disagreement with Heisenberg's new scientific work, but also contempt for the man, a deep resentment that the founder of quantum physics had helped the Nazis try to build an atomic bomb. Those in attendance were stunned by the drama.

Here was a firsthand example of the ethical considerations that sometimes arise in science, and the controversy and bitterness Heisenberg aroused among many fellow scientists when he chose to remain in Germany during World War II. Such controversy haunted the rest of his life, although he continued to receive medals, honorary degrees, and invitations to speak. For most of the war, Heisenberg was a professor of physics at the University of Berlin and director of the Kaiser Wilhelm Institute for Physics. Many German scientists, and especially Jews such as Einstein, Lise Meitner, and Hans Krebs, had already fled Germany or been expelled. Other scientists, such as Max Planck and Max von Laue, remained in Germany but opposed the Nazi regime and managed to avoid doing war-related research. Not Heisenberg. He led the German effort to build an atomic bomb as well as other wartime scientific work.

Heisenberg's motives and thinking during the war will probably always remain hidden in a cloud of uncertainty. There is evidence that early in the war, Heisenberg miscalculated the critical mass of uranium needed to build the bomb, estimating a much larger amount than was actually required. To purify that enormous amount of the right isotope of uranium would be far beyond the resources of the German government, and the German atomic bomb effort came to a halt (unknown to the Allies).

One can wonder what Heisenberg would have done if the needed resources had been given to him. Heisenberg did not support the brutal regime of the Nazis. As the Austrian American physicist Victor Weisskopf later said, "It must have driven him to utter despair and depression that his beloved country had fallen so deeply into the abyss of crime, blood, and murder." While Heisenberg could have emigrated, he could have joined the anti-Nazi underground movement, or he could have retired from public life, he was a loyal German. He was not a hero. And, having won a Nobel, he was too prominent to retire. What of emigration, the first possibility? Here, Heisenberg had a terrible choice. In her biography, titled *Inner Exile,* Heisenberg's wife, Elisabeth, says that her husband felt that leaving Germany would have spared his reputation but nothing else. In emigrating, "he would have abandoned his friends and his students, his family in general, physics . . . just to save himself. It was a thought that he could not bear."

In his memoir written many years later, Heisenberg described his thoughts about building the bomb and his moral considerations at the time. Whether that description is honest and accurate, or whether an attempt to rehabilitate his reputation, we do not know.

> Toward the end of 1941 our "uranium club" had, by and large, grasped the physical problems involved in the technical exploitation of atomic energy. . . . As for the production of uranium 235, we knew of no feasible methods that could have yielded significant quantities in Germany under war conditions. In short, though we knew that atom bombs could now be produced, in principle and by what precise methods, we overestimated the technical effort involved. Hence we were happily able to give

the authorities an absolutely honest account of the latest development, and yet feel certain that no serious attempt to construct atom bombs would be made in Germany. . . .

Nevertheless, we all sensed that we had ventured onto highly dangerous ground, and I would occasionally have long discussions particularly with Carl Friedrich von Weizsäcker, Karl Wirtz, Johannes Jensen and Friedrich Houtermans as to whether we were doing the right thing. . . .

I tried to put myself into [the Americans'] position.

The psychological situation of American physicists, and particularly of those who have emigrated from Germany and who have been received so hospitably, is completely different from ours. They must all be firmly convinced that they are fighting for a just cause. But is the use of an atom bomb, by which hundreds of thousands of civilians will be killed instantly, warrantable even in defense of a just cause? Can we really apply the old maxim that the ends sanctify the means? In other words, are we entitled to build atom bombs for a good cause but not for a bad one? And if we take that view—which has unfortunately prevailed throughout history—who decides which cause is good and which bad? It's easy enough to see that Hitler's cause is a very bad one, but is the Americans' good in every respect? Must we not judge it, too, according to the means by which it is pursued? Of course, even the good fight invariably involves some bad means, but is there not a point beyond which we cannot go under any circumstances? During the last century people tried to set a limit to the use of evil means through pacts and conventions. But in the present war these conventions are probably being ignored by Hitler no less than by his opponents. All in all, I think we may take it that even American physicists are not too keen on building atom bombs. But they could, of course, be spurred on by the fear that we may be doing so.

The physicist Lise Meitner (1878–1968), who did pioneering work in nuclear physics that laid some of the theoretical foundations for the bomb, had this to say: "[Before the war], one could love one's work and not always be tormented by the fear of the ghastly and malevolent things that people might do with beautiful scientific findings."

The case of Werner Heisenberg illustrates the moral and ethical dilemmas scientists must sometimes wrestle with. And, in the case of the atomic bomb, the American physicists at Los Alamos, who actually built the bomb, had to go through some of the same soul-searching that Heisenberg describes above. Although we may view science itself as a sanitized pursuit of knowledge, without ethical or moral intent, the products of science, such as knowledge of the uranium atom or knowledge of how genes may be edited, can be used for good or for ill. Science itself does not have values. Science does not have notions of right or wrong. It is we human beings who have values.

In the laboratory with their test tubes and voltmeters, or at their desks with pencil and paper, scientists are usually the first to understand the potential uses—good and bad—of what they have discovered. Being on the front lines, they should be obligated to consider the consequences of their discoveries. But they cannot bear that burden alone. They have a responsibility to share their findings with their larger society so that judgments and decisions can be made by that society, at least by the policymakers and leaders. In such ethical and moral decisions, scientists have no more competence or authority than other citizens. But no less.

MAGDALENA LENDA

A mild day in May 2017, and thirty-four-year-old ecologist Magdalena Lenda was doing what she loved best. She was out in the field, immersed in grasslands and plants and trees, working on a research project to measure the impact of invasive species on native plants near Cracow, in southern Poland, her own native habitat. She bent down to identify a particular plant, then recorded the data by pencil in her notebook. Identification is her specialty. "My father taught me how to identify plants," she recalls. "It happens automatically in my head. I don't need any book or any guide to identify species of plants. It's automatic. I just look at the plant, and I know what it is."

Dr. Lenda's father sold fish for a living, but in his free time he took young Magdalena on trips to the countryside, especially to abandoned orchards. He taught his daughter what plants and fruits she could eat. In part, such instruction was a utility. At that time, during the communist period in Poland, food in shops was limited. But in addition, the rural excursions delighted the girl. "Identifying species and observing nature was our entertainment," she says. "This is how I got interested in nature."

For the project that day, Magdalena and a half-dozen colleagues were testing the conjecture that two invasive species acting simultaneously can totally wipe out native species, essentially ending biodiversity in the region. They were working in an abandoned field of 1 hectare in size (about 2.5 acres). Their full plan involved twenty such fields. Surrounding them, in the distance, one could see other abandoned fields and also the lush rolling hills of active farms growing rye, potatoes, and apples, appearing as neat stripes of greens and taupes and chartreuse. Dressed in her usual attire for fieldwork, Magdalena wore sports shoes, jeans, a T-shirt and sweater, and a backpack with water and coffee, sweets, and sandwiches. A six-hour day in the field is normal, but when she was a PhD student, she sometimes went for eleven hours straight.

The researchers marked off a dozen circular areas of study with 2-meter-wide toy hula hoops. The resulting plots were to serve as the units of measurement. Enlisted as the bad guys, the invasive plants were goldenrod, indigenous to North America, and walnut trees, indigenous to Asia. The scientists chose plots so that some contained only native plants, some contained goldenrod as well, some contained walnut (from the walnut tree canopy above), and some contained both goldenrod and walnut. Various members of the team then took soil samples, while others measured the richness and coverage of native plants in each plot.

Dr. Lenda led the group. She enjoys working with other people. "My idea of collaboration is building long-lasting relationships and research groups. The more diverse the people with different ideas, points of view, and cultures, the better for research." She also loves simply hanging out with people.

When she's not out in the field, she may be cooking for friends, or dancing Argentine tango at clubs in Cracow.

Over the next few months, after careful measurements and a thorough statistical analysis, Dr. Lenda's team arrived at the somewhat surprising conclusion that the negative impact on native plant diversity is much *reduced* when both goldenrod and walnut are present, compared to the damage done by either alone. Their findings were published in the journal *Diversity and Distributions.* In the discussion part of the paper, Lenda and her colleagues suggest that future studies of invasive plants should take into account the interaction of two or more alien species.

Six years and many more research projects later, Lenda says that she is a person motivated by making changes. And she does not accept received wisdom. If someone says that a bad situation is impossible to change, she rolls up her sleeves and gets to work. "Most scientists say that there is nothing we can do to stop invasive species," she says. "But I say no. If you observe invasive species in a national park, decreasing 90 percent of biodiversity, this is a disaster, and we should take some actions. That's why I work with invasive species. I want to change my society."

In her own way, Magdalena Lenda is an activist as well as a scientist. Among all the scientists we've recently talked to, it is likely that Lenda most bridges the distance between science and public policy, especially environmental policy. Her scientific concerns are closely connected to her interests in our human relationship to nature.

—

Magdalena Lenda's mother, a biologist who worked in epidemiology in government institutions, as well as her father, undoubtedly influenced her career. In 2005, Magdalena received a bachelor's degree in geology with a specialization in nature conservation from Jagiellonian University (in Cracow), founded in 1364 and one of the oldest universities in Europe. The following year, Magdalena had a profound experience that confirmed her decision to become an ecologist. She was walking in an abandoned field covered by walnuts. After an hour of carefully paying attention to what was growing and what was not, she realized that land abandonment is a trigger for invasion by nonnative species.

It was an "aha" moment. "This was super interesting for me," she recalls many years later. "This is when I definitely determined to become a scientist, especially someone who studies invasive species."

The concept of conservation biology as a field of scientific study is recent. One might trace its beginnings to 1978, with a conference titled the First International Conference on Research in Conservation Biology, held at the University of California, San Diego, in La Jolla, California. The conference was motivated by an increasing concern for tropical deforestation, disappearing species, and eroding genetic diversity within species.

For her PhD thesis, Magdalena's research took an unexpected detour from her fieldwork: the analysis of how the internet has increased the distribution of invasive plants around the world. In a recent paper titled "Misinformation, internet honey trading, and beekeepers drive a plant invasion," Lenda showed that scientifically unsupported claims about the health benefits of goldenrod honey have spread faster and farther with the rise of social media and the internet. Such widespread misinformation created a market demand for goldenrod honey even in countries where goldenrod is absent. In turn, beekeepers in those countries planted goldenrod, which outcompetes native plant species that are food for pollinators, ants, and birds.

Lenda continues to expand her research areas in unusual ways. She is currently studying for a second PhD in psychology, on the subject of the interplay between human psychology and nature conservation. Her recent research in this area explores how contact with local nature, such as watching birds from one's window, identifying species of local animals and plants, and backyard gardening—all viable options for people who cannot travel far from their homes—has benefits for mental health. The positive effect of nature on mental health has previously been documented. But Lenda's study focuses on experiences with nature that are nearby and readily available, as during the lockdowns in the recent COVID pandemic, when many people were housebound. Paradoxically, the actual research for this local nature project and for the goldenrod honey study was about as far from nature as you can get. In both cases, the "field" was not the sweet-smelling meadows and grasslands that Lenda loves, but the glassy screen of a computer. Without leaving her office, Lenda surveyed and analyzed hundreds of reports available on the internet. Her research was a study of studies.

Unlike the research of some scientists, single-mindedly focused on one phenomenon for an entire career, Dr. Lenda's work has nimbly skipped from one area to another. As an example of yet another direction, some years ago a student in her group studied biological light-emitting diodes implanted in trees, with the ultimate goal of using such phosphorescent plants to light city walkways in place of energy-consuming electric lights. For Lenda, the constant redirection of her professional work and exploration of new projects is necessary to keep her engaged. "By always finding new projects, I never get bored with science," she says. "It is very important for a scientist not to burn out. Because if you are bored with your work, you are not using your creativity."

Despite the large range of topics and research methods, Lenda's work remains rooted in nature. Every creative person—scientists, writers, painters, musicians—draws from some deep well of inspiration or some galvanizing experience. For Lenda, that deep well is nature. "On the one hand, it is my job," she says, "but on the other hand, it is the kind of activity that allows me to clear my mind. Contact with nature was always very calming for me. Whenever I had some problems as a teenager, I went to the forest or to some meadows. I just like being in nature. If I can also get work and earn money from contact with nature, it is my dream job."

Magdalena is as adventurous in her personal life as she is in her professional life. She takes tango classes twice a week and, on weekends, goes with friends to milonga, a club or nice restaurant with tango dancing and music. "Tango dancing is fun, and it allows me to focus on totally different activities," she says. "It is something different from science. It is more emotional. We exchange partners every song. Sometimes there are many people who would like to dance with us."

She also loves to experiment with cooking different kinds of ethnic and international food. "I have never learned how to cook Polish food," she says, laughing. Her friends, in the many countries she has worked, "showed me how to cook amazing food that comes from Papua New Guinea, from Nigeria, from Nepal, from Thailand, and from Malaysia. I am a big fan of experimental cooking. I always try to follow the recipes, but also to make some innovations." Magdalena enjoys finding out about cultures other than her own. "When I travel, I am trying to meet local people and to eat local food with them and to experience activities with them, like cooking or going into the field with them so that they can show me their local nature and to talk with them about their culture."

On weekends, when she is not dancing, Magdalena often cooks for friends.

"Feeding people is what you can do to show love," she says. "You can never show people that you like them more than cooking for them. If the cooking is done with pleasure and love and honesty, it is something you would like to give to your friends or partner."

It would seem that Magdalena has compartmentalized her life: science during the week, tango and cooking and friends on the weekends. But she is like many scientists (and musical composers and CEOs of companies) in that when she is immersed in a project, she thinks about it nonstop. She's driven to know answers. "The topics that I am working on, they are always fascinating," she says. "You know, if you have something that is fascinating, it doesn't matter if it is the weekend, or the evening, or even the night. You just want to find out more about this thing." One is reminded of Walt Whitman's line when he realized that he was destined to become a poet: "Never more shall I escape."

The balance of work life with the rest of life is a matter of personal temperament, for scientists just as for nonscientists.

Neuroscientist Lace Riggs says that on weekends she cannot think of anything more fun than her scientific work. By contrast, physicist Dorota Grabowska strongly prefers not to work on weekends, needing time to wind down. Magdalena fits somewhere in between.

Doing science, doing what she loves, is in some ways a privilege, her "dream job," as Magdalena says. Especially pure science, without direct application, might be considered a luxury as well as a dream. Pure science is usually supported by national governments, ultimately depending on the taxes paid by citizens of those nations. But how can societies justify such expenditures, without any direct benefit? Dr. Lenda has an answer to that question. "Basic science has value like art. For me, knowledge is like an art. Not all knowledge can be applied, but it is important to have this knowledge to know how the world works. It is the same with art. We will not die if we don't know who van Gogh is. However, we would be poorer as humans if we never saw van Gogh's paintings."

Although she is not involved with politics, Lenda worries about whether the Polish government will continue to support the art of basic science and even applied science. Scientists in other countries have similar worries. In the last decade or so, we have witnessed a growing anti-science movement worldwide and a mistrust of scientists as part of the "elite establishment." "It can happen that we scientists will not be able to do what we want to do," she says, "because the government will oppose our work. I may be forced to search for a job in another country. But I don't want to emigrate. I want to serve my society here as long as I can."

Chapter VII

PATTERNS OF SCIENTIFIC DISCOVERY

In one of the most remarkable narratives of scientific discovery, German pharmacologist Otto Loewi got the idea for testing how nerves communicate with one another in a dream. The night before Easter Sunday of 1921, he awoke, turned on the light, jotted down his idea, and immediately went to the lab, where he performed a simple experiment on a frog heart.

At the time, it was well-known that signals travel down the spindly filaments of a nerve in the form of electricity. What was not known was how nerves conveyed their electrical impulses to other nerves, muscles, and organs. In short, how do nerves talk to the rest of the body? Most biologists believed that such communication was also electrical—that tiny currents flowed from the nerves to heart muscle or thyroid gland or the waiting antennae of other nerves.

Loewi's late night experiment was not only simple. It was elegant. First, he isolated the live hearts of two frogs and removed the nerves from the second heart. (The hearts were placed in solutions that kept them alive.) He then stimulated the vagus nerve of the first heart. The vagus nerve slows down the functions of organs, and the frog heart's rate of throbbing decreased, as expected. After a few moments, Loewi poured liquid from a tube inserted into the first heart into the second, nerveless heart. *It also slowed.* Its beating diminished, just as if its vagus nerve had been roused. The experiment showed, without a doubt, that a stimulated nerve releases some chemical substance that then activates other organs. Loewi had discovered the existence of neurotransmitters, such as acetylcholine and adrenaline. (See chapter 3 for more on neurotransmitters.)

Einstein had his thought experiments, nuclear physicist Ernest Rutherford had his "damn fool" ideas, and biologist Barbara McClintock recognized meaning in the patterns of leaves of corn. But few scientists have reported receiving their great ideas in a dream.

How do scientists arrive at their discoveries? Which discoveries were accidental and which were purposeful? From examining the discoveries of a number of physicists, biologists, and chemists, we have been able to group scientific discoveries into several different categories. Of course, some discoveries bridge more than one category below.

The Accident. Here scientists discover something they were not looking for. This category breaks down into two subcategories: accidental discoveries whose significance the scientists did not understand at the time, and accidental discoveries whose significance was immediately appreciated. A good example of the first was the discovery of the cosmic background radiation by Arno Penzias and Robert Wilson in 1965. (See the profile of John Mather for a discussion of the cosmic background radiation.) Penzias and Wilson, both excellent experimentalists, had no explanation for the residual hiss in their radio antenna and did not understand its meaning. Indeed, at one time they thought that the hiss was static caused by the droppings of pigeons on their antenna. Not far away, in the H-shaped redbrick Palmer Laboratory at Princeton University, Robert Dicke and his group had already calculated that such a pervasive background radio signal, at a frequency corresponding to a temperature of a few degrees above absolute zero, would be evidence for the Big Bang. In fact, Dicke and his team were in the process of building their own detector when they heard news of Penzias and Wilson's unexplained hiss. "They've got it," Dicke said to his team, undoubtedly with mixed feelings on the realization that they had been scooped. (Penzias and Wilson, but not Dicke, were awarded a Nobel for their discovery, an essential underpinning of the Big Bang model.)

An example of the second subcategory was Alexander Fleming's discovery of penicillin in 1928. Although Fleming was completely surprised to discover a white fluff growing on his culture of staphylococci and the dissolving of the staph near the white mold, he instantly realized that he had stumbled upon an antibacterial agent.

Principles First. Here scientists begin with a philosophical or logical principle and explore the consequences of that principle, sometimes initially unaware of the precise problem to be solved. The application of the foundational principle can, in fact, define both the problem to be solved and its solution. The premier example of this rarefied category is Einstein's work on special relativity, in which he began with the symmetry principle that all frames of reference moving at constant velocity relative to one another are equivalent. (For example, if you are in a train moving at constant speed and throw a ball, its trajectory will appear exactly the same to you as if you are sitting in the train station and throw the same ball.) In the process of working out the consequences of his starting principle, applied to light, a form of electromagnetic radiation, Einstein discovered that our concept of time had to be reconceived.

Principles Last. Here the scientists engage in concentrated work to explain a particular experimental result and ultimately recognize that a new, fundamental theoretical idea is needed. Max Planck's discovery of the quantum in 1900 illustrates this category. The German physicist was trying to use the methods of thermodynamics and statistical mechanics to justify his own ad-hoc formula for the universal spectrum of radiation inside a black container (so-called black-body radiation). To do so, he had to assume that the energies of the vibrating atoms of the container were not continuous, but came in discrete lumps, or quanta. Here began the new field of quantum physics.

The Timely Clue. Here the scientist is confronted with an important clue while struggling with a recognized problem. An example of this category of discovery was Niels Bohr's creation of the first quantum model of the atom in 1913. While trying out different models of the atom, in which various parameters of the electrons' orbits about the atomic nucleus were quantized using Planck's new quantum constant, Bohr was shown Johann Balmer's 1885 empirical formula for the frequencies of light emitted by hydrogen atoms. The key clue was that each emitted frequency depended on one number subtracted from another (like $\frac{1}{4} - \frac{1}{9}$, and $\frac{1}{4} - \frac{1}{16}$; the numbers subtracted from each other were actually the reciprocals of squares of integers: $(\frac{1}{2})^2$ and $(\frac{1}{3})^2$ and $(\frac{1}{4})^2$, etc.). Bohr had never before seen the formula. "As soon as I saw Balmer's formula," Bohr later recalled, "the whole thing was immediately clear to me." Bohr then hypothesized that the electrons in atoms emitted photons (particles of light) when passing from one energy level to another, the energy of the photon being the difference in the energies of the two levels.

Another illustration of the timely clue was Barbara McClintock's discovery in the late 1940s that genes could move around on chromosomes and, in the process, alter controls, commands, and the storage of information. For some years, McClintock had been pondering how genes, including pigment-controlling genes responsible for yellow leaf streaks, were turning on and off during the growth of a single corn plant. Variations in streakiness appeared to occur in a regular way, not in the random fashion you would expect from mutations. Carefully examining her plants one day in 1946, McClintock noticed that these controlled mutations came in pairs. For example, if one section of a corn leaf had a higher than average number of streaks, a neighboring section had a lower than average number. Here was a big clue. Since each of the sections of the leaf originated from a different progenitor cell, it appeared as if one of the progenitor cells had given something to the other during cell division.

McClintock recalled that when she first observed this twin-sector phenomenon, it was "so striking that I dropped everything, without knowing—but I felt sure that I would be able to find out what it was that one cell gained and the other cell lost." This observation led first to the idea that control elements (genes) were passed between sister chromatids (threadlike filaments into which a chromosome divides in a cell division) and later that control elements could change positions on individual chromosomes.

In both of these examples, the scientist needed a good understanding and vision to recognize the importance of the clue. A clue that comes before its time is of no use. The remarkable fit of the opposite coasts of Africa and South America, like adjacent pieces of a jigsaw puzzle, was known for centuries but never recognized as a clue to the geography of the ancient Earth until in 1912 Alfred Wegener proposed his theory of continental drift: the radical idea that landmasses could move horizontally across the surface of the Earth. Even then, many geophysicists remained skeptical, deeming the Earth too rigid to allow such motions. Not until the 1960s was the concept widely accepted, when measurements of upwelling and spreading of the ocean bed supported a theory of plate tectonics, according to which the continents were floating on less rigid material in the Earth's core.

The story of continental drift illustrates the fact that clues, sometimes very strong clues, are often ignored until a comprehensive and compelling theory comes along to explain them. This is especially true when scientific clues are anomalous—that is, they violate current thinking. Another example is the so-called flatness problem. As early as 1961, physicist Robert Dicke pointed out that it was very odd that the gravitational energy of the cosmos at large is almost equal to its kinetic energy of expansion, a near equality almost impossible to explain in terms of the standard Big Bang model. Dicke's comment was essentially ignored by the scientific community until the inflationary universe model of Alan Guth and others in 1979. That new theory proposed an exponentially fast expansion of the infant universe, far faster than the leisurely expansion rate of the standard Big Bang model, and it naturally explained the near equality of cosmic gravitational and kinetic energies. (The inflation model has since been supported by other evidence.)

Likewise, the adaptation-of-organism problem, in which many animals seem exquisitely adapted to their environments. For example, the fat-storing humps of camels, which allow the rest of their bodies to keep cool in the desert, and the tall necks of giraffes, allowing them to eat from the tall trees in their environments. Before the mid-nineteenth century, most naturalists and many others took these adaptations to be evidence of Grand Design. But after Darwin's theory of natural selection, they were seen to follow naturally from evolutionary forces.

Analogy. Here the scientist notices a pattern in a previous phenomenon and reasons that such a pattern might apply in a new situation, like noticing that all the even and odd house numbers are on opposite sides of the street in Boston and looking for the same pattern in Chicago. An example of this kind of discovery was Hans Krebs's 1937 uncovering of the citric acid cycle, later known as the Krebs cycle. Krebs and other biochemists were trying to discover what chain of chemical reactions was responsible for the combination of oxygen with carbohydrates and fats to produce energy in living organisms, starting with citric acid. Several years earlier, Krebs had made the first discovery of a cyclic process in biochemistry, in which a chemical called ornithine is changed to citrulline, which is changed to arginine, and then back to ornithine, ready to begin the cycle again. (Along the way, the toxic molecule ammonia is converted to urea, which is then removed from the body in urination.) *Krebs had cycles on his mind.* The scientist searched for the missing chemical steps that would regenerate citric acid at the end of the metabolic process and return to the starting point of the cycle. As he wrote in his memoir, "In visualizing the cycle mechanism, it was of major relevance that five years earlier, I had been concerned with the first metabolic cycle to be discovered, the ornithine cycle of urea synthesis."

The Mathematical Imperative. Here the scientist, while exploring the mathematical world, is led to a discovery about the physical world. A prime illustration of this type of discovery was Paul Dirac's formulation in 1928 of the equation describing the electron. The requirement that such an equation embrace both relativity and the new quantum physics necessitated a particular mathematical structure. In following the narrow path of this mathematical landscape and its internal logical consistency, Dirac was led to his equation, which actually predicted the existence of a new class of elementary particles called antiparticles, never before seen. As Dirac said in an interview in 1963, his discovery of the correct equation describing the electron "came out just from playing with the equations rather than trying to introduce the right physical ideas. A great deal of my work is just playing with equations and seeing what they give. . . . I think it's a peculiarity of myself that I like to play about with equations, just looking for beautiful mathematical relations which maybe don't have any physical meaning at all. Sometimes they do."

New Tools. Here the availability of new instruments opens up opportunities for breakthroughs. An example of this type of discovery occurred in 1929, when American astronomer Edwin Hubble observed that the velocities of recession of galaxies was approximately proportional to their distances from us, leading to the conclusion that the universe is expanding. Other scientists were attempting to measure the distances to a group of fast-receding galaxies. Hubble's advantage lay in his access to the relatively new 100-inch Hooker Telescope on Mount Wilson, in California, at that time the largest telescope in the world.

Another example was the first discovery of a black hole, made possible by the Uhuru X-ray satellite in 1970, the first telescope able to detect X-rays from outer space. Uhuru was also one of the first telescopes launched into space aboard an orbiting satellite. Uhuru revealed many previously unknown astronomical objects that emit X-rays.

The Long Haul. Here the steady, often slow, incremental work on a recognized problem over a long period of time eventually leads to a discovery. An example was Max Perutz's unraveling of the three-dimensional structure of hemoglobin, one of the first protein structures to be fully understood. Perutz and his team worked on the problem for twenty-two years, from 1938 to 1960, painstakingly producing and analyzing hundreds of X-ray diffraction photographs and refining their experimental techniques along the way.

Another example also involving the improvement of experimental techniques over many years was the discovery of gravitational waves in 2015 by an instrument called LIGO (Laser Interferometer Gravitational-Wave Observatory).

Physicists worked on LIGO for forty years from its first conception to its eventual success. Einstein's theory of gravity predicted the existence of gravitational waves, and theorists had calculated the likely strength of such waves from distant astrophysical sources such as the collision of black holes. These ripples in space have tiny amplitude—equivalent to the thickness of a hair at the distance of Alpha Centauri—so exceedingly sensitive equipment is needed to detect them, requiring several generations of new technologies.

Although there is clearly a large range of processes in scientific discovery, some common patterns emerge. Most discoveries involve a synthesis, bringing together different strands of information or ideas and connecting them. To formulate his quantum model for the atom, Bohr combined Rutherford's nuclear model of the atom, Planck's idea of quantized energy levels, and Einstein's idea of the photon (quantum particle of light).

Another pattern that occurs in many but not all scientific discoveries is the following sequence of events: research and hard work, leading to the prepared mind; being stuck on the problem; finally, a shift in thinking or perspective. The shift in thinking does not necessarily come suddenly, in a "eureka" moment. In fact, eureka moments, of the kind described by Heisenberg in chapter 1, are rare.

But, regardless of the time each step takes, the general sequence of steps toward new understanding seems to be common. German physicist Lise Meitner's discovery of nuclear fission followed this pattern. So did Barbara McClintock's work on the mutations of maize, as did James Watson and Francis Crick's work on the structure of DNA. Watson and Crick, with the benefit of Rosalind Franklin's X-ray photographs, already knew that DNA had a spiral-like structure, and they knew the chemical compounds making up that structure. So they had prepared minds. But they were having trouble chemically fitting the base pairs together, the steps of the DNA ladder. Then chemist Jerry Donohue told them that the dangling hydrogen atoms on the nitrogen bases should be in different locations. That fact provided a change in thinking about how the bases could fit together. Donohue was the only person Watson and Crick thanked in their paper.

The prepared mind is critical. We know of no examples of major scientific discoveries in the twentieth century made by untrained amateurs. Even when the finding is accidental, it requires a prepared mind, as in the case of Fleming's discovery of penicillin. The Scottish scientist had been working on antibacterial agents since his 1908 medical school thesis. Being stuck also seems important in many discoveries. This frustrated mental state, combined with a prepared mind, can catalyze the creative imagination. These patterns are probably universal to the creative process in general and occur in the arts as well as in the sciences.

As discussed in chapter 1, there is no single scientific personality type. Most if not all scientists, as children, show a curiosity about the world, a wonder, and an independence of mind. When nuclear physicist Lise Meitner was a child, her grandmother cautioned her that she should not sew on the Sabbath or else the heavens would come tumbling down. The little girl decided to do an experiment. She lightly touched her needle to some embroidery and looked up. Nothing happened. Then she took a stitch, waited, looked up. Again, nothing. Finally, Lise decided that her grandmother had been mistaken and went happily about her sewing.

GOVIND SWARUP

During the summer months of the late 1930s, Govind Swarup and his brothers used to sleep on the roof of their house and look up at the stars. It was a large and fine house, built from the considerable profits of a spinning mill constructed and run by Govind's great-grandfather. Although there are no photographs or diaries from those nocturnal summers on the roof, we might imagine the scene: a small town in northern India named Thakurdwara, population about two thousand at the time, in the state of Uttar Pradesh, lying in a lush green part of the country. The Himalayas can be seen in the distance, only dim shadows at this hour. Since nightfall, the temperature has cooled to the low seventies Fahrenheit. As the boys try to sleep, they might hear the hooting of Bengal eagle owls, nesting in the teak and sal trees, which fill the air with a sweet and spicy aroma. The sals, especially, are considered sacred in the Hindu tradition and associated with the deity Vishnu.

Many years later, Govind recalled those summers lying on the rooftop: "I looked at the stars in the sky and wondered what they were, why many scintillate and some not at all. My teachers could not tell me about the stars. Astronomy was not taught in any school or college during the British rule of India—the syllabus was possibly prescribed to provide the required [Indian] manpower for ruling the country."

Govind described his father, Ram, as "very enterprising." In 1931, Ram built one of the first cinemas in Delhi, showing Hindi films. He also developed a 200-acre farm in the village of Dhouti, where Govind spent many vacations. In 1941, Govind's father moved the family to the city of Moradabad. There, Govind studied in the Hindu High School from grades eight to ten. During that period, the boy occasionally rode an elephant to school. The animal had been bought by his father to work on the farm during the wet season, but when the elephant's services were not needed, it was available for the boy.

Govind remembered that in August 1942, at the age of thirteen, he joined the other students of the school in a protest march, shouting "Quit India" in support of the call made by Gandhi to end British colonial rule. "The soldiers fired on us peaceful students, though only in the air, but that episode enraged me a great deal." Although young Govind didn't join any other public protests, he was inspired by Gandhi and later read the Indian leader's writings in his weekly magazine *Harijan.*

In high school, Govind read an article by the poet Mahadevi Verma about the Milky Way galaxy. "I found it fascinating," recalls Govind, "and read more about [the Milky Way] in an astronomy book in the school library. That was my first lesson in astronomy! . . . I became keenly interested in science."

Such were Govind Swarup's beginnings in science. Born into a prosperous family, his father, grandfather, and great-grandfather were all businessmen, not scientists. Govind developed his love of science on his own. Eventually he would become one of India's greatest astronomers, a pioneer in the new field of radio astronomy, conceiving and building novel telescopes still used around the world.

Govind died in 2020 at the age of ninety-one, renowned in the community of professional astronomers but largely unknown beyond. His obituaries in the Indian press christened him "the Father of Radio Astronomy in India."

Radio telescopes, of the kind Govind Swarup designed, detect radio waves from outer space. Radio waves are part of the electromagnetic spectrum, traveling waves of electrical and magnetic energy. These waves come in different lengths, the distance between successive crests. Visible light consists of electromagnetic waves ranging in length from about 0.000045 centimeters (blue light) to about 0.000075 centimeters (red light). Ultraviolet and X-rays, invisible to the eye, have much shorter wavelengths. Radio waves, also invisible, are much longer.

Radio astronomy began in 1931, two years after Swarup was born, when the electrical engineer Karl Jansky (1905–1950) accidentally discovered radio emissions coming from the direction of the Milky Way, the gauzy band of light arcing across the sky on a clear night, the plane of our galaxy. At the time, Jansky was working with a radio antenna at the Bell Telephone Lab in New Jersey. We now know that many unusual and energetic objects in outer space, such as highly dense, rapidly rotating burned-out stars called pulsars, emit their radiation primarily in the radio band of the electromagnetic spectrum. Thus, radio telescopes and detectors have greatly added to our knowledge of the cosmos.

The antennae, often shaped as giant dishes, need to be large for several reasons. First, as for any telescope, the larger the area of the receiver, the more radiation comes in, and for very distant and faint objects, you want as much radiation as possible. Secondly, the wavelengths of radio waves are much longer than the wavelengths of visible light. Larger dishes are needed to compensate for these longer wavelengths. Here's why. The degree of detail that can be seen through the lens (or dish) of a telescope, called its resolving power, is proportional to the number of complete wavelengths spanning the lens. (A complete wavelength consists of one crest and one trough.) Longer wavelengths of light require larger lenses to collect the same number of wavelengths. Indeed, modern radio telescopes, including the ones designed by Swarup, often have many dishes separated by several miles. When the received signals of these many dishes are brought together and properly synchronized, the entire array acts as if it were a single dish several miles in diameter. Such telescopes can resolve details as small as about one-thousandth of a degree, depending on the exact wavelength of the radio wave, equivalent to discerning a penny at a distance of about 3 miles.

In July 1944, Govind's father sent him to the city of Allahabad to study mathematics, physics, and chemistry. It would have been most natural for Govind to go into business, following the tradition of his father and grandfather. And perhaps that is what Govind's father ultimately had in mind. But Govind's imagination was ignited by pure science, and his childhood experience lying on rooftops and pondering the stars must have been a strong influence.

In 1950, Swarup joined the National Physical Laboratory (NPL). There, under the supervision of his former teacher at Allahabad, Govind built an instrument to use radio emissions to study magnetic materials in the lab. He recalls that to build the apparatus, he used surplus components from World War II radar equipment, which were lying in the basement of NPL, a resourcefulness that he was to rely on again and again in a country with limited scientific parts and supplies.

In 1953, Govind went to Australia to work in the Division of Radiophysics of the Commonwealth Scientific and Industrial Research Organisation, where many of the new discoveries in radio astronomy were being made. This was Govind's first trip out of India. Here, he became familiar with the new developments and techniques in radio astronomy. The young scientist, still in his midtwenties, worked with a number of groups in building radio telescopes. Govind and his group made daily observations of the radio emission from the Sun. As he recalls, "The Sun was then a very enigmatic celestial radio source, one of the few known at that time." Govind asked, How did energy get from the Sun's relatively cool interior to its extremely hot outer corona? (It is now believed that turbulent motions and magnetic fields carry energy there.)

In 1956, Govind married Bina. As he recalled, "That was the start of a partnership that continues to this day. Ours was an arranged marriage; her father found me through a newspaper advertisement." Govind and Bina were to have two children, a daughter, Anju, and a son, Vipin.

In 1959, as a graduate student at Stanford University in the United States, Swarup helped to develop radio interferometry—a technique for combining the signals from an extended array of radio antennae, allowing them to act in concert as one giant antenna. (Radio waves arriving at separated dishes will arrive at slightly different times, having traveled different distances. When their signals are combined at a central location, their respective crests and troughs won't quite line up. Adjusting for this misalignment effectively adds the signals from the dishes and pins down the location of the source.) Swarup's method is still used today in radio telescopes all over the world, as well as in other applications such as the synchronization of atomic clocks miles apart and in transmitters in space.

After receiving his PhD, Swarup, now well-known in the world of astronomy, returned to India and joined the Tata Institute of Fundamental Research in Mumbai. There he began creating his own group of young astronomers. He also founded and directed the Tata Institute's National Centre for Radio Astrophysics, in the sprawling city of Pune.

As with John Mather and other leaders of big experimental and observational science teams, Swarup possessed people skills in addition to scientific skills, allowing him to congenially lead the groups of scientists and engineers needed to build large radio telescopes. Despite being a respected leader, he was a humble and gentle man. One photograph shows him shoulder to shoulder with a dozen other men literally hoisting a 45-meter-wide radio antenna up to its mount by hand-operated winches. A colleague at the Tata Institute described Swarup as "always full of innovative ideas. . . . He had no hesitation in discussing or bouncing his thoughts and ideas off his colleagues and students, often in the corridors or canteen or any other place where he might meet them. Some may even say that his ideas evolved quite rapidly and sometimes were difficult to keep up with. But the sparkle and curiosity in his eyes on a wide variety of topics, discussing what is most exciting in a field with intense energy and enthusiasm, was not only a delight to see but very inspiring too, especially to young students. . . . Besides his enthusiasm, vision, and energy, what touched me from those early days was the humility and simplicity of the man, operating in a very egalitarian ambience from a modest office. . . . He never availed of any special privileges for himself."

A long associate of Swarup and a core member of his team, S. Ananthakrishnan, recalled that "his ready and infectious smile hid a sharp mind. An innovator with high skills, he was extremely hardworking and focused on his goal. Most of all, he enjoyed the company of young bright minds and could argue with them on equal terms on scientific problems; it invigorated everyone. This ambience was different from what one finds in many academic circles today in India." Another former student says that Swarup "created an environment such that young scientists and engineers never hesitated to go up to his office and tell him their true feelings and opinions."

In 1963 Swarup and his team built India's first radio telescope, at Kalyan, near Bombay. It was an array of thirty-two parabolic dishes, each 6 feet in diameter.

Two other Swarup telescopes were world-class and of unusual design. They were remarkable because they took advantage of the benefits of the Indian economy—cheap construction—and avoided the need for electronic components that would be hard to manufacture in that country. Swarup also cleverly used topography. His Ooty Radio Telescope was built at a location with latitude 11 degrees north, but on a hillside with a slope of 11 degrees, so that it was aligned parallel to the Earth's axis.

Traditional telescopes making a long observation have to rotate in two directions—vertical and horizontal—to track an object that moves across the sky as the Earth itself rotates. But because of its special alignment parallel to the Earth's axis, the Ooty telescope had to rotate in only one direction.

Swarup's second world-class telescope was the Giant Metrewave Radio Telescope (GMRT), which consists of thirty antennas, each with a diameter of 45 meters (148 feet), spread over a region of 25 kilometers (16 miles). Swarup minimized the cost of the GMRT by using stainless steel mesh instead of solid metal for the large dishes. This innovation in materials also made the telescope more maneuverable. The GMRT, which was proposed and completed by Swarup between 1987 and 1997, is still regarded as the world's largest radio telescope at certain radio wavelengths.

Among many other discoveries, the GMRT helped scientists identify remnants of the Ophiuchus Supercluster explosion, produced by a massive black hole at the center of a large galaxy in the Ophiuchus constellation. The gravitational energy released by matter falling into the black hole was cataclysmically transformed into giant jets of high-energy particles streaming away from the galaxy. This violent event, thought to have occurred at least 250 million years ago, is considered to be the biggest explosion in the history of the universe.

At age sixty-five, Swarup retired from the Tata Institute. The year was 1994. But he still had work to do. As he recalled, "At that time, I was very concerned about the lack of quality education in physics, mathematics, chemistry, and biology in India. I dream of an India where we have world-class universities." Swarup was the prime mover behind creating the Indian Institutes of Science Education and Research, offering a five-year course integrating education and research. "I consider this a bigger achievement than building the GMRT," Swarup told an interviewer.

A photograph of Govind at the 2013 meeting of the Astronomical Society of India shows him in center front. By this time, he is nearly bald, with a fringe of gray hair. He is wearing his usual dress: an open-collar white shirt and dark trousers, sandals, and a warm, humble smile on his face. Beside and around him are about two hundred astronomers, young and older, men and women, all inspired by this leader of radio astronomy in India, and the world.

Govind Swarup, more than most of the scientists profiled in this book, went beyond his research career to help advance his entire country, through science education. He keenly felt an obligation toward his society. The social obligations and responsibilities of scientists will be discussed more in the next chapter.

Chapter VIII

THE ETHICS OF SCIENCE AND THE RESPONSIBILITIES OF SCIENTISTS

On July 3, 1945, ten German scientists who had worked on Germany's nuclear program were interned by the Allies at a country mansion called Farm Hall, in Godmanchester, England, about 18 miles northwest of Cambridge. The purpose of incarcerating the physicists was to find out how close Nazi Germany had been to building an atomic bomb, and possibly also to keep them from falling into the hands of the Russians. The scientists included Otto Hahn, who in 1938 had discovered that uranium could fission and who had received a Nobel Prize in 1944; Werner Heisenberg, one of the creators of quantum mechanics and also a Nobel Prize winner in 1932; and Carl Friedrich von Weizsäcker, who made important contributions to the physics of energy production in stars.

A month later, on the afternoon of August 6, 1945, the German scientists learned that an atomic bomb had been dropped on Hiroshima. At first, they didn't believe the news, as they had previously concluded that the construction of such a weapon would be prohibitively expensive. Then, as more information began trickling in, they decided it was true. According to secret tape recordings at Farm Hall and subsequent recollections of Heisenberg's, Otto Hahn felt enormous guilt that "his greatest scientific discovery now bears the taint of unimaginable horror." There followed a remarkable conversation between Heisenberg and Weizsäcker about the ethics of science and responsibilities of scientists.

"The word 'guilt' does not really apply," Heisenberg said to Weizsäcker, "even though all of us were links in the causal chain that led to this great tragedy. Otto Hahn and all of us have merely played our part in the development of modern science. . . . We know from experience that it can lead to good or to evil." Then Weizsäcker responded.

> There will, of course, be quite a few who will contend that science has gone far enough. . . . They may, of course, be right, but all those who think like them fail to grasp that, in the modern world, man's life has come to depend on the development of science. If we were to turn our backs on the continuous extension of knowledge, the number of people inhabiting the earth in the fairly near future would have to be cut down radically. . . .
>
> For the present, the development of science is a vital need of all mankind, so that any individual contributing toward it cannot be called guilty. Our task, now as in the past, is to guide this development toward the right ends, to extend the benefits of knowledge to all mankind, not to prevent the development itself. Hence the correct question is: What can the individual scientist do to help in this task; what are the precise obligations of the scientific research worker?

Heisenberg then says that "the individual tackling a scientific or technical task, however important, must nevertheless try to think of the broader issues."

And Weizsäcker again: "In that case, if [the scientist] wants to act for the best and not just leave it at noble thoughts, he will probably have to play a more deliberate part in public life, try to have a greater say in public affairs. Perhaps we should welcome this trend, for inasmuch as scientific and technical advances serve the good of society, those responsible for them will be given a greater say than they currently enjoy. Obviously, this does not mean that physicists or technicians could make better political decisions than the politicians themselves. But their scientific work has taught them to be objective and factual, and, what is more important, to keep the wider context in view."

The ethics of science and the responsibilities of scientists do not have simple formulations or prescriptions. Heisenberg said that modern science can lead to good or to evil. But sometimes it is not easy to define the good. For example, is it morally justified to build a weapon to kill people, if by killing a few we can save the lives of many? Is it morally justified to alter the DNA of human embryos in order to make the resulting human beings smarter or more athletic? Should a scientist stop working on a fundamental research problem, like the way memory is stored in the brain, if she thinks it might possibly lead to harmful applications?

Our view is that science and the technology resulting from science do not have values in themselves. It is we human beings who possess values. And we should employ those values in how we use science and technology. The good referred to by Heisenberg probably meant—as it does for many people—increasing the well-being, happiness, and quality of life of the largest number of people. And the bad diminishes that well-being. We further suggest that scientists, as citizens of their society, have a responsibility to ensure that their discoveries and innovations are used for good and not for bad. Such a responsibility, of course, means that scientists will have to take some time away from their lab benches and equations to engage with the public and with policymakers. We also suggest that scientists, as citizens of the world, share a responsibility to help relieve the world's economic inequalities, including the relative lack of access to energy, food, health care, and technology in the Global South. As Weizsäcker said, scientists are not policymakers, nor do they have the required skills. But their special expertise and evidence-based thinking should be a resource for policymakers to improve the lives of everyone. And, as we live in a scientific and technological age, buffeted by rapid developments in biotechnology, artificial intelligence, and many other areas, scientists have a responsibility to educate the public in scientific matters. Policymakers may often be motivated by self-interest, but ultimately, in democratic societies, they must answer to the public.

In our view, the areas of science and technology now posing the greatest ethical dilemmas and challenges are artificial intelligence, biotechnology and synthetic biology, advanced medical procedures, and climate change. Artificial intelligence is already revolutionizing many aspects of our lives, including health care, banking, transportation, information exchange, and even warfare. New computer programs are able to learn things by themselves, as well as utilize vast data banks, and will someday become fully autonomous, operating without human input. Biotechnology—the manipulation of biological processes and the DNA of microorganisms to produce novel products—is already being used to create such things as batteries, drugs, improved fertilizer and other agricultural products, and new engineering devices. This rapidly developing field began with the understanding of the structure of DNA in the 1950s. Advanced medical procedures include the ability to edit the DNA of human embryos, new procedures for extending the lives of permanently bedridden patients, and the rapid sequencing and analysis of a person's full DNA, revealing psychological tendencies, origins of personality, and potential illnesses.

We discussed in chapter 1 that there is not always a clear demarcation between pure and applied science. Discoveries in pure science often later lead to applications, such as Einstein's theories of relativity in 1905 and 1915 (now essential to GPS systems), the invention of the transistor in 1947 (used in computers, electronic equipment, and telecommunication devices), the unraveling of the structure of DNA in 1953 (now used to identify pathogens, in the treatment of cancer, and many other applications), the discovery of mRNA in 1961 (the basis for the COVID vaccines), and the discovery of carbon nanotubes in 1991 (used to make plastics with enhanced electrical conductivity and for delivering drugs and nerve cell regeneration).

Today we live in a world more dependent on technology than ever before and ever more vulnerable to its failures or misdirection. To be at ease in this fast-changing world and to be effective citizens, everyone needs at least a basic grasp of science's concepts and discoveries. Scientific education and communication aren't just for scientists. Obviously pandemics, climate change, and AI have been at the forefront of our minds recently, but policies on health, energy, and the environment all have a scientific dimension. To grasp their essence isn't so difficult: most of us appreciate music even if we can't compose or perform it. Likewise, the key ideas of science can be accessed and enjoyed by almost everyone. The technicalities may be daunting, but these are less important for most of us and can be left to the specialists.

The Earth has existed for 45 million centuries, but this is the first when one species—ours—has the planet's future in its hands. We have entered a new geological era, the so-called Anthropocene, where the main threats come from us and not from nature. If public debate on relevant ethics and policy is to rise above mere sloganeering, everyone needs to have enough of a feel for science to avoid becoming bamboozled by propaganda and bad statistics. The need for proper debate will become even more acute in the future, as the pressures on the environment and those resulting from misdirected technology get more diverse and threatening. In this respect, one of the most frightening outcomes from the recent populist movements across the globe has been the death of factual and evidence-based thinking. In today's post-truth era, there is little agreement on what defines reliable sources. The internet, with all of its virtues, has abetted the rapid diffusion of misinformation and disinformation.

Only geniuses or cranks venture to tackle head-on the broad scientific mega-challenges of biotechnology, fundamental physics, and artificial intelligence. The prudent and realistic strategy is to focus on a bite-sized problem that will be at least a step toward its solution. On the other hand, the occupational risk of this approach is that scientists may forget that the narrow problems they're tackling are mainly worthwhile only insofar as they are steps toward answering some big questions. That's why it is good for scientists to engage with general audiences. In fact, when one discusses the great unknowns, there is less of a gap between the specialist and the audience. When even the experts haven't much clue, they are in a sense in the same position as the public; they're perhaps confused at a deeper level, but that is all. Even if we scientists explain ourselves badly, we benefit from exposure to general audiences who focus on the big questions and remind us of how much we still don't know. Robert Wilson, the radio engineer who made the serendipitous discovery of the cosmic background radiation—which clinched the case for a Big Bang—said that he himself didn't fully appreciate the import of his momentous work until he read an article headlined "The Afterglow of Creation." Good journalists offer a breadth and critical perspective that can atrophy in professional scientists through overspecialization, so good journalistic work can benefit specialists as well as the wider public.

The interconnectedness of today's world, by virtue of global trade, the internet, and global challenges such as climate change, requires scientists to engage with the international community, not only their own societies. We depend on elaborate networks: electric power grids, air traffic control, international finance, just-in-time delivery. Unless these are highly resilient, their manifest benefits could be outweighed by catastrophic (albeit rare) breakdowns cascading through the system. Pandemics can spread at the speed of jet aircraft, causing maximum havoc in the shambolic but burgeoning megacities of the developing world. Social media can spread psychic contagion—rumors and panic—literally at the speed of light. The issues impel us to plan internationally. For example, whether or not a pandemic gets a global grip may hinge on how quickly a Vietnamese poultry farmer can report any strange sickness. And many of these challenges—energy and climate change, for instance—involve multi-decade timescales, plainly far outside the concern and comfort zone of most politicians. Consequently, politicians need the best in-house scientific advice in forming their policies. But, more than that, these issues should be part of a wide public debate, and such debate must be leveraged by scientific citizens—engaging from all political perspectives with the media and with a public attuned to the scope and limit of science. Scientists can act through campaigning groups, via blogging and journalism, or through political activity. There is a role for national academies too. Politicians, informed by their scientific advisors, should aim to lift long-term global issues higher on the political agenda, where they are habitually trumped by the urgent and parochial.

Scientists should present policy options based on a consensus of expert opinion, but if they engage in advocacy they should recognize that on the economic, social, and ethical aspects of any policy they speak as citizens and not as experts. But they all have a responsibility. You would be a poor parent if you didn't care what happened to your children in adulthood, even though you may have little control over them. Likewise, scientists shouldn't be indifferent to the fruits of their ideas and creations. They should try to foster benign spin-offs—commercial or otherwise. They should resist, so far as they can, dubious or threatening applications of their work and alert politicians when appropriate. We need to foster a culture of responsible innovation, especially in fields like biotech and advanced AI.

Of course, scientists have special obligations over and above their responsibility as citizens. There are obviously ethical obligations confronting scientific research itself: avoiding experiments that have even the tiniest risk of leading to catastrophe and respecting a code of ethics when research involves animals or human subjects. But less tractable issues arise when research has ramifications beyond the laboratory and has a potential social, economic, and ethical impact that concerns all citizens—or when it reveals a serious but still-unappreciated threat. And scientists are in a better position than nonscientists to understand the potential physical dangers of their work.

There is another, perhaps more subtle issue. We would argue that scientists, and the journalists who write about science, have a responsibility to be careful in the *interpretation* of scientific results, especially when presenting those results to the public. In chapter 4, we mentioned that in January 2023, the World Health Organization (WHO) stated that "no level of alcohol consumption is safe for our health." That statement was repeated in *Time* magazine and other publications for the general public. However, the WHO headline might have been an overstatement of the actual results. It was based on a July 14, 2022, article in the medical journal *The Lancet.* A careful reading of that article shows that: (1) rather than finding that no level of alcohol consumption is safe, the researchers could not determine the threshold below which consumption is safe, and (2) a small amount of alcohol might have positive benefits for older adults, outweighing the negative effects. This case illustrates the fact that scientists and science writers must concern themselves not only with the actual scientific data but with the interpretation and presentation of that data. The interface between the science itself and the presentation of results to the public is a layered hierarchy, winding its way from the lab bench and statistical tests to the scientists' analysis and interpretation of their results to the publication of those results in scientific journals to the interpretation and presentation by science journalists writing for popular publications. Care must be exercised at every step of this hierarchy.

Care must also be taken, by both scientists and science journalists, not to exaggerate or sensationalize scientific findings. There are too many examples of popular articles (written both by scientists themselves and science journalists) that claim a crisis in their field, when only modest revisions of current understanding are warranted. It is far easier to get such exclamatory articles published than those that report the more ordinary progress of science. Overly dramatic headlines contribute to the mistrust of scientists and their institutions.

One can highlight some fine exemplars of citizen scientists from the past: for instance, the atomic scientists who developed the first nuclear weapons during World War II. Fate had assigned them a pivotal role in history. Many of them—men such as Joseph Rotblat, Hans Bethe, Rudolf Peierls, and John Simpson—returned with relief to peacetime academic pursuits. But for them the ivory tower wasn't a sanctuary. They continued not just as academics but as engaged citizens—promoting efforts to control the power they had helped unleash through national academies, the Pugwash movement, and other public forums. They were the alchemists of their time, possessors of secret specialized knowledge. Nuclear physics was twentieth-century science. But there are now other technologies—bio, cyber, and AI—that have implications just as momentous as nuclear weapons. In contrast to the atomic scientists, those engaged with the new challenges span almost all the sciences, are broadly international, and work in the commercial sector as well as in academia and government. Their findings and concerns need to inform planning and policy. To echo Weizsäcker, how is this best done?

Direct ties forged with politicians and senior officials can help—and links with NGOs and the private sector too. But experts who've served as government advisors have often had frustratingly little influence. Politicians are, however, influenced by their inboxes, and by the press. Scientists can sometimes achieve more as outsiders and activists, leveraging their message via widely read books, campaigning groups, blogging and journalism, or—albeit via a variety of perspectives—through political activity. If their voices are echoed and amplified by a wide public and by the media, long-term global causes will rise on the political agenda. Rachel Carson and Carl Sagan, for instance, were both preeminent in their generation as exemplars of the concerned scientist, and they had immense influence through their writings and speeches. And that was before the age of social media and tweets.

A special responsibility resides with scientists in academia or self-employed entrepreneurs. They have more freedom to engage in public debate than those in government service or in industry. And those of us who are academics have a special privilege to influence successive generations of students. We should try to sensitize them to the issues that will confront them in their careers. Indeed, polls show, unsurprisingly, that young people are more engaged and anxious about long-term and global issues than those in earlier generations. Student involvement in, for instance, effective altruism (doing good better) and long-terminism (taking moral responsibility for future generations of people) campaigns is burgeoning.

The ethics behind such campaigns is not straightforward. Even if all human lives have equal intrinsic value, don't we have special obligations to those closest to us? Are there other desirable goals beyond happiness? And to what extent should we sacrifice present benefits for the long-term future, given that in our fast-changing world we can't confidently predict the needs and preferences of future generations?

New technology and applied science, rather than pure science, have the most immediate impact on our lives. New technology has a checkered history.

For example, it has always been used, in part, to build weapons of destruction. But its principal stated purpose is to improve the lives of people. As discussed below, that goal was truer in the past than today, when monetary incentives are so enormous.

One of our heroes in the history of science is Francis Bacon. In his *New Atlantis* (1627) Bacon envisioned a kingdom of science and technology, most of it unheard-of at the time, that included living chambers where the air is treated for the preservation of health, where new methods are developed for the perfection of agriculture, glass lenses are invented for seeing objects far off, and sound technologies are studied to produce hearing aids. In this utopian kingdom, called Salomon's House, three Benefactors were charged with sifting through the experiments of all the house scientists "to draw out of them things of use and practice for man's life and knowledge."

Soon after Bacon, the development of technology became part of a major Western intellectual theme called "progress." The movement centered around the notion that human beings were inevitably advancing to a higher plane—socially, politically, intellectually, morally, and scientifically. In France, Nicolas de Condorcet's *Sketch of the Intellectual Progress of Mankind* (1795) proposed the concept of "infinite perfectibility" of humankind. In this grand idea of progress, intellectual progress was represented most notably by the discoveries of Isaac Newton and his sweeping laws of motion and gravity. Material progress was nowhere better symbolized than in James Watt's remarkable steam engine, the centerpiece of the Industrial Revolution. Power looms, for example, enabled textile workers to perform at ten or more times their previous rates and promised to raise the standard of living and relieve the exploitation of factory workers. Concern for the human condition was central in these developments.

Another of our heroes of science is Benjamin Franklin (1706–1790). Franklin refused to patent his many inventions for private profit because he felt that citizens should serve their societies "freely and generously." Among other things, Franklin invented methods for cleaning the streets of London and a new kind of stove, still in use today, "for the better warming of rooms, and at the same time saving fuel." For Franklin and many other technologists of the day, the human being always came first.

According to Leo Marx and other historians of technology, at least two developments in the mid-nineteenth century changed the nature and perception of technology. First, some areas of technology began to evolve from the individual-oriented mechanic arts, like glassblowing and woodworking, to large, depersonalized systems, like the railroad. Secondly, these vast technological systems became hugely more profitable than any previous technology in the history of the world, offering great personal wealth to their creators. (Think Cornelius Vanderbilt, Andrew Carnegie, John D. Rockefeller, and, more recently,

Steve Jobs, Elon Musk, Larry Page, Sergey Brin, Mustafa Suleyman, and Jeff Bezos.) No longer was technology a humanistic activity, with its principal purpose to improve the quality of life. Technology went from a means to humanitarian progress to an end in itself, and an instrument of capitalism. Progress was technology, and technology was progress. According to this new perception of technology, if a new optical fiber can quadruple the transmission of data, then we should develop it. If a new plastic has twice the strength-to-weight ratio of the older variety, we should produce it. If a new automobile can accelerate at twice the rate of an older model, we should build it. All without a necessary regard for improving the quality of life. Technology as a thing in itself.

In the nineteenth century, some of the new technologies were the railroad and the telegraph. Today, some of the new technologies are the computer, the internet, and their many manifestations and applications such as ChatGPT, Meta, Instagram, Twitter/X, Snapchat, Spotify, Uptake, Groupon, TikTok, *Subway Surfers,* Reddit, Venmo, PictureThis, BirdsEye, Merlin, Tinder, Flipp, Twitch, Caffeine, Smule, Webex, Discord, Peanut, and literally thousands more. How many of them were invented to improve the human condition?

Let's now consider in a bit more detail the major areas of science and technology that pose the greatest challenges in the near future: advanced medical procedures; artificial intelligence; biotechnology and synthetic biology; climate change, energy, and food; and space technology.

—

Medical. Throughout the history of medicine, people have recoiled at innovations that seemed to go against nature, including vaccination, transfusions, artificial insemination, organ transplants, and in-vitro fertilization. The fact that these innovations are unexceptionable today (aside from recent resistance in some quarters to the COVID-19 vaccine) is a reminder that squeamishness at the new is not a reliable guide to what eventually becomes ethically defensible and broadly acceptable. The gulf between what future advances in medical science may enable us to do, and what is prudent or ethical to do, will shift, widen, and, in many cases, be difficult to cope with.

Most people distinguish between an intervention that would remove something harmful, which they welcome, and one that would enhance what we already have, which they fear. Whether or not this difference is morally significant (or even, in many cases, meaningful), the actual prospect of genetic enhancement of humans is, perhaps fortunately, remote. A few genetic diseases, including Huntington's, are caused by a single gene that could be snipped out by the CRISPR/Cas9 gene-editing technique. But most, such as schizophrenia or a susceptibility to Alzheimer's or cancer, are the product of hundreds or thousands of genes, each tweaking the probability of a person having the disease by a tiny amount. This collaboration of genes may be even truer of traits and talents such as height, intelligence, and personality. Only when the DNA and trait profiles of many millions of people are available will it become possible (using pattern-recognition systems aided by AI) to identify desirable combinations of genes. Not until this can be done—and genes can be confidently synthesized without inducing new downsides—will designer babies become conceivable.

But there are genuine concerns about where advancing genetics could take us. (Indeed, there is widespread reluctance to properly discuss the issue, because it seems redolent of the now anathematized eugenic movements in the first half of the twentieth century.) Research on aging exemplifies such concerns. There is an obvious incentive to pursue research with the aim of prolonging our healthy lifespan. Altos has set up labs aiming to address these problems in California (in the San Francisco Bay Area and in San Diego) and in Cambridge (UK), funded by some U.S. billionaires—people who, when young, aspired to be rich, and having achieved riches, want to be young again. That's not so easy! Will the benefits be modest and incremental? Or is aging a disease that can be held at bay or even eliminated? Dramatic life extension, if it proved possible at all, would initially be the privilege of a wealthy elite—a fundamental new kind of inequality. But if life extension became widespread, it would plainly be a real wild card in population projections, with huge social ramifications (such as multi-generational families? a later menopause? etc.).

In addition to the distinction between curing and enhancing, many people draw a line between genetic manipulations whose effects are restricted to individuals' own bodily tissues and those that reach into the eggs or sperm and are passed down to their progeny. Indeed, manipulation of the egg and sperm cells of other species forces us to think hard about our ethical position. For instance, there has been an attempt in parts of Brazil and other areas to sterilize and thereby reduce—and even wipe out—the species of mosquito that spreads the Zika and dengue viruses. The trials recorded a 90 percent reduction in local populations of the species. Is it wrong to play God in this way? Similar techniques are being proposed that could preserve the unique ecology of the Galápagos Islands by eliminating invasive species such as black rats.

If the relevant technologies continue to advance, there seems a real long-term prospect that human beings—their brains and their bodies—could be enhanced via genetic and cyborg modification. (For example, computer chips in our brains that communicate directly with the internet, or alteration of the lenses of our eyes to see X-rays.) Moreover, this future evolution—a kind of secular intelligent design leading even to a new species—might take only centuries, in contrast to the thousands of centuries needed for Darwinian evolution.

Genetic manipulations are by no means the only ethical challenges that will sharpen as biomedical science advances. We will also face acute dilemmas about treating those at the beginning and those at the end of their lives. Everyone treasures the prospect of living out more healthy years, while most people dread the prospect of being kept alive in pain or with severe disability or dementia. "Assisted dying" or "voluntary euthanasia" is now legalized, with safeguards, in several European countries, and several U.S. states. In the UK, public opinion is 80 percent in favor of legalization—under closely circumscribed conditions. Professional medical opinion is shifting toward acceptance and seems now evenly balanced; one can even find archbishops on both sides of the argument. Likewise, the ability to treat premature babies can be miraculous, but it might also mean saving children who will never flourish, laying out an ethical minefield.

AI. Already, artificial intelligence can cope better than humans with data-rich, fast-changing networks, such as traffic flow and electric grids. The Chinese may develop an efficiently planned economy that Marx only could have dreamed of. And AI can help science too—with protein folding, drug development, and maybe settling whether the physicists' string theory can really describe our universe.

Clearly, machines will take over much of manufacturing and retail distribution. They can supplement, if not replace, many white-collar jobs: accounting, legal research, computer coding, medical diagnostics, and even surgery. In contrast, skilled service-sector jobs—plumbing and gardening, for instance—that require non-routine interactions with the external world will be among the hardest to automate.

The implications for our society are already ambivalent. If we're sentenced to a term in prison, recommended for surgery, or even given a poor credit rating, we would expect the reasons to be accessible to us and contestable by us. If such decisions were entirely delegated to an algorithm, we would be entitled to feel uneasy, even if presented with compelling evidence that, on average, the machines make better decisions than the humans they have usurped.

ChatGPT and its more powerful successors will surely confront us, in more extreme and insidious forms, with the downsides of existing social media: fake news, fake photos and videos, and unmoderated extremist diatribes. Remember that AI deploys its speed and memory by processing text. It doesn't have any concept of the real world.

These scenarios are sufficiently near term that we need to plan for them and adjust to them. But what about the longer-term prospects? These are murkier, and there is no consensus among experts on the speed of advance in machine intelligence—and indeed on what the limits to AI might be. It seems plausible that an AI linked to the internet could clean up on the stock market by analyzing far more data far faster than any human. To some extent this is what quantitative hedge funds are already doing. But for interactions with humans, or even with the complex and fast-changing environment encountered by a driverless car, processing power is not enough; computers would need sensors that enable them to see and hear as well as humans do and the software to process and interpret what the sensors relay.

Regarding AI in general, we should be concerned, at least in the short term, not so much by the sci-fi scenario of a takeover by super-intelligence as by the risk that the world's economic and social infrastructure will become dependent on networks whose failure could cause a societal breakdown that cascades globally.

We would argue that some kind of international regulation is needed for AI safety. Innovations need to be tested before wide deployment, like the rigorous testing of drugs that precedes government approval and release. But regulation is a special challenge in a sector of the economy dominated by a few vast multinational conglomerates. Just as they can move between jurisdictions to evade fair taxation, such conglomerates could evade regulations of AI. We applaud the recent AI Act adopted by the European Parliament. Scientists and technologists have a responsibility for advising policymakers, at both the national and international level, on reasonable guidelines and regulations for AI.

Regarding regulations, biotechnology and AI in particular are advancing so rapidly that they will be extremely difficult to regulate, especially since most of the advances are occurring in private companies. Governments are typically too slow, and too bureaucratic, to respond quickly enough to new developments. Here, Eric Schmidt, former CEO of Google, has offered the interesting suggestion that such fields be monitored by incentivized private testing companies, competing with one another in a marketplace environment. The most successful of such testing companies would be financially rewarded by governments and ultimately taxpayers.

We suggest that the particular scientists involved with AI also have a responsibility to help ensure that AI be used in the interests of humanity rather than only to put money in the hands of commercial companies like Microsoft and Google. Many, if not most, commercial companies are motivated almost completely by making money rather than improving the well-being of people. Such a capitalistic motivation can and does produce innovations that improve the quality of life, but not always. A recent case with OpenAI, the makers of ChatGPT, illustrates the challenge. The founders of OpenAI realized that AI is too powerful to be completely controlled by profit-seeking companies and organized their company around a dual structure, with a nonprofit board of advisors overseeing the commercial side of OpenAI. Indeed, the mission of the nonprofit piece is that AI should "act in the best interests of humanity." In November 2023, the company fired its chief executive, Sam Altman, apparently because he was not sufficiently reining in the for-profit part of OpenAI. Altman was immediately courted by Microsoft. Within a matter of days, he was rehired at OpenAI, the leaders of which had evidently decided that they could not resist the siren call of capitalism.

—

Biotechnology. In the early days of recombinant DNA (gene splicing) in the 1970s, there was concern about unintended consequences, and a moratorium was imposed following the recommendations of a conference of experts held at the Asilomar Conference Center, California, in February 1975. This moratorium soon came to seem unduly cautious, but that doesn't mean that it was unwise at the time, since the level of risk was then genuinely uncertain. The moratorium showed that an international group of leading scientists could agree to a self-denying regulation and influence the research community powerfully enough to ensure that it was implemented. There have recently been moves to control the still more powerful techniques of synthetic biology, which can create organisms from the ground up by synthesizing a new genome bit by bit. A voluntary consensus is harder to achieve today. The academic community is far larger, and competition (enhanced by commercial pressures) is more intense.

And, of course, research on viruses, such as the precursors of COVID-19, highlights dilemmas that have proved incendiary. In 2011 researchers at the Erasmus University in the Netherlands and at the University of Wisconsin in the United States showed it was surprisingly easy to make the H5N1 influenza virus both more virulent and more transmissible—defying the evolutionary dynamic that ordinarily trades one of these features against the other (since a virus that quickly kills its host can no longer use that host to spread itself). These Faustian gain-of-function experiments were justified as a way to stay one step ahead of natural mutations, but they could be used with evil intent. The U.S. federal government banned these experiments in 2014, but for reasons that seemed somewhat unclear, relaxed the rules three years later. This is another area where scientists have a responsibility to inform policymakers about the current technologies and ways to contain them.

Climate Change, Energy, and Food. Over most of history, the benefits we garner from the natural world have seemed an inexhaustible resource, and the worst terrors humans confronted—floods, earthquakes, and diseases—came from nature. But not anymore. We're now deep in what some have called the Anthropocene Era, in which human activity is having a significant effect on our climate and the global system of living things. The human population, now over 8 billion, makes collective demands on energy and resources that aren't sustainable without new technology and resources, and possibly not even then.

Looming over the world this century is the threat of human-induced climate change. This threat is potentially a global fever, in some ways resembling a slow-motion version of COVID-19. For instance, both crises aggravate the level of inequality within and between nations. Those in the megacities of the developing world can't isolate from rogue viruses; their medical care is minimal; and they're less likely to have access to vaccines. Likewise, it's those countries and the poorest people in them that will suffer most from global warming and the effects on food production and water supplies.

Climate change and environmental degradation may well, later this century, have global consequences that are even graver than pandemics—and longer term (indeed irreversible). But a potential slow-motion catastrophe doesn't engage the public and its politicians. Our predicament resembles that of the proverbial frog in a pot of boiling water—contented in a warming tank until it's too late to save itself. We're well aware of the threats, but we fail to prioritize countermeasures because their worst impact stretches beyond the time horizon of political and investment decisions. Politicians recognize a duty to prepare for floods, terrorist acts, and other risks that are likely to materialize in the short term and are localized within their own domain. But they have minimal incentive to address longer-term threats that aren't likely to occur while they're still in office—and that are global rather than local. Scientists and other concerned citizens have an obligation to take the long view.

A benign scenario for the global environment would leave a large fraction of each region in a biodiverse and natural state, preserving at least the present level of wildlife and vegetation. Energy would be derived from clean sources. Production would be as pollution-free as possible. Indeed, some processes could be carried out in space. There could be a network of underground fiber and cables to carry energy and information. Water could, when necessary, be pumped great distances to avoid droughts in any region. If nuclear energy is widespread, the challenge of safe disposal of radioactive waste will need to have been solved. Buildings would be long-lived; if they had a short lifetime they would be designed so that the material is recyclable or reusable. And new technologies would be developed (indeed such innovation is already taking place) to draw carbon dioxide out of the air. A modern version of Bacon's Salomon's House might encompass such a view.

But there is a contrasting scenario—a catastrophic one. We could have a world still burning fossil fuels with a different climate. Large areas of our coastal cities will need to be evacuated because of rising sea levels, placing great immigration pressures on countries with more habitable climates. High-priority construction projects worldwide would include seawalls and embankments to protect against rising sea level and more violent storms, but these constructions might be futile in certain parts of the world.

Choices made before 2050 will determine which of these scenarios plays out. There's a balance to be struck between mitigating climate change and adapting to it. And there are other ethical questions crucial to this debate. How much should we sacrifice now to ensure that the world is no worse when our grandchildren grow old? How much subsidy should be transferred from the rich world, whose fossil-fuel emissions have mostly caused the problem, to the developing nations? How much should we incentivize clean energy? Should we gamble that our successors may devise a technical fix that will rectify our bad habits now? On all these choices there's as yet minimal consensus, still less effective action. But policies and investment priorities must be influenced by climate change projections.

Climate change and the needed reduction in fossil-fuel burning is, of course, closely related to demands for energy. And on the energy front, science can offer a win-win road map to a low-carbon future. Nations should accelerate research and development into all forms of low-carbon energy generation and into other technologies where parallel progress is essential. Especially needed here are new ideas for energy storage, such as advanced batteries, compressed air, pumped storage, underground storage, and hydrogen technology, which are all crucial if we're dependent on unsteady generation via sun or wind. Integrated long-range low-loss grids should also be considered—to bring solar energy from North Africa and Spain to less-sunny northern Europe, and to smooth over the peak consumptions (normally around seven p.m.) in different time zones by east-west transmission lines—across North America and maybe also all the way along the Belt and Road to China.

Achieving such a result will require vision, commitment, and public-private investment on the same scale as the building of railways in the nineteenth century. Complete transformation of the world's energy and transport infrastructure is a massive project that will inevitably take several decades.

Wind and solar energy generation are both being widely deployed, though there are niches that can be filled by geothermal, hydro, and tidal power. And what about nuclear power? Despite the ambivalence about widespread nuclear energy, it's surely worthwhile to boost R & D into fourth-generation nuclear power—especially small modular reactors—which could be more flexible and safer than existing reactors. Indeed, they could also be much cheaper if all follow a standard design. The number of nuclear power stations being built in Western countries has plummeted in the last twenty years. Current designs date back to the 1960s and have become immensely costly because of extra equipment required to enhance safety. Countries like France that now depend heavily on nuclear energy will need to replace existing power stations; replacements are scheduled for the 2030s (and indeed several in the UK have faults that require earlier decommissioning).

Nuclear fusion, the process that powers the Sun, still beckons as an inexhaustible source of energy. Attempts to harness this power have been pursued since the 1950s. There have been some false dawns, and the history is of receding horizons; commercial fusion power still seems at least thirty years away. Most prototypes deploy magnetic forces to confine gas at a temperature of millions of degrees—hotter than the center of the Sun. Despite its cost and the formidable challenges it poses, the potential payoff of fusion energy is so great that it is surely worth continuing these developments. In December 2022, a fusion reactor at the Lawrence Livermore National Laboratory in the United States, in which fuel pellets were compressed and then zapped by lasers, was for the first time able to produce more energy than was delivered to the pellet—though still one hundred times less than the overall energy consumption of the system.

There's a broader and even more compelling commercial motivation for prioritizing clean energy research in technically advanced countries: Under business as usual, the main increases in annual CO_2 emissions will come from the countries in Asia and Africa, which can't reach acceptable living standards without generating more power than they do today. Not only will their per capita energy needs rise, unlike ours in the West, but they will collectively harbor a billion more people by 2050—4 billion rather than 3. Changing the trajectory of CO_2 emissions from these nations is crucial. They must be economically and technically enabled to leapfrog to clean energy rather than building coal-fired power stations (as they have transitioned to smartphones without ever building landlines). Technically advanced countries have some responsibility to help countries in the less advanced Global South make such an energy transition.

Climate change and energy needs bring up the urgent issue of food security. World food production needs to double by 2050, not only to cope with the rise in population but also to ensure that all those in the Global South (the source of the main population growth in coming decades) become as well-nourished as most people in Europe and North America. A mega-challenge to young idealistic scientists and engineers is to push forward efforts to provide sustainable supplies of food and energy for the entire world.

Thanks to the Green Revolution, food production has doubled in the last fifty years, but a further doubling is more problematic. There will be constraints on energy, on the quantity of fertile land, and on the supply of water. These challenges will require further improved agriculture—low-till, water-conserving, and GM crops—together with greater efforts to reduce waste (via refrigeration, for instance) and improve irrigation. We need modes of farming that can produce crops efficiently in a changing climate and avoid encroaching on natural forests. The buzz-phrase is "sustainable intensification." There will be consequent pressure to enhance the yield from the oceans without allowing overfishing to drive species to extinction. There will certainly need to be changes in the typical Western diet; for instance, we can't all consume as much beef as present-day Americans.

Space Technology. Its impetus, from World War II onward, came from the defense sector: intercontinental ballistic missiles (ICBMs) were deployed in the thousands during the Cold War period. But we depend every day on more benign uses of space—for communications, weather forecasting, GPS and satnav, and environmental monitoring. And, of course, astronomers would extol all that's been learned about the wider cosmos: robotic vehicles have trundled over the surface of Mars; spacecraft have traveled to the edges of our solar system—to Pluto and beyond—beaming back pictures of varied and distinctive worlds. Moreover, rockets have launched into orbit a succession of increasingly powerful telescopes that have offered a clearer view of the remote universe—revealing exotic cosmic objects that can't be detected from the ground because their main emission is in infrared, ultraviolet, X-rays, or gamma rays, which are absorbed by the Earth's atmosphere.

Until now, scientific discovery has been the main driver for sending probes to the Moon and beyond. Future waves of exploration are likely to have a more commercial bent, perhaps motivated by the quest for raw materials, such as precious metals from asteroids, that will profit private companies back on Earth. There are clear ethical and regulatory questions to be asked about what the boundaries of those activities should be. Space law is no longer a purely academic topic. On a moral basis, do humans, as private individuals, corporations, or members of some favored nation, possess the right to alter or even destroy the landscape of other celestial objects? Can real estate on the Moon or Mars be owned? And can governments be prevented from using space as a battleground?

A crewed mission to Mars, including provisions and the rocketry for a return trip, could cost the United States hundreds of billions of dollars. One reason for the high cost is that NASA, in response to public attitudes, has developed a safety culture greater than that during the Apollo era. This concern with safety reflects the national trauma and consequent program delays that followed the space shuttle disasters in 1986 and 2003, each of which killed the people on board. The shuttle had 135 launches altogether, so the two crashes represented a failure rate below 2 percent. It would be unrealistic to expect a probability this low for the failure of a return trip to Mars.

There will certainly be thrill seekers and adventurers who would willingly accept far higher risks—and even sign up for a one-way trip. A crucial contrast between the Apollo era and today is the emergence of a strong private space-technology sector, which embraces human spaceflight. Private-sector companies like SpaceX and Blue Origin are now competitive with NASA, so high-risk, cut-price trips to Mars, bankrolled by billionaires and private sponsors, could be crewed by willing volunteers. But there should be no illusions. These space trips should be sold for what they actually are: a dangerous sport, rather than space tourism, with its connotations of a routine, low-risk activity.

The pioneer space explorers will be ill adapted to their new habitat, so they will have a more compelling incentive than those of us on Earth to redesign themselves. They'll harness the super-powerful genetic and cyborg technologies that will be developed in coming decades. These technologies will, one hopes, be heavily regulated on Earth, on prudential and ethical grounds. But settlers on Mars will be far beyond the clutches of the regulators. The modification of their progeny to adapt to alien environments may be the first step toward divergence into a new species, what one might call *Homo techno,* as discussed in chapter 2.

As mentioned earlier, we *Homo sapiens* are surely not the culmination of evolution of the genus *Homo,* and our brains can't conceive what might replace us. Indeed, we are already evolving into *Homo techno,* with increasing use of digital technology integrated into our bodies. Such transformations may occur within a century—within the lifetime of children already born, and therefore certainly within the planning horizon on which scientists, ethicists, or policymakers should focus.

—

Despite the various concerns we have raised, there are grounds for optimism. For most people in most nations, there's never been a better time to be alive. The innovations driving economic advance—information technology and biotech—can boost the developing as well as the developed world. Creativity in science and the arts is nourished by a wider range of influences—and is accessible to more people worldwide—than in the past. We're becoming embedded in a cyberspace that can link anyone, anywhere, to all the world's information and culture and to most other people on the planet. More should be done to assess and then minimize the risks and challenges we've discussed here. But though we live under their shadow, there seems no scientific impediment to achieving a sustainable and secure world, where all enjoy a prosperous lifestyle while being environmentally sustainable and involving lower demands on energy.

We can be technological optimists, even though our technology may need redirection. And that redirection must be guided by values that science itself can't provide. As we have argued, science and technology do not have values in themselves. It is we human beings who have values. And it is the responsibility of scientists and technologists, both as specialists and as citizens of the world, to help advise policymakers and governments. There are certainly difficulties. Politicians look to their own voters and the next election. Stockholders expect a payoff in the short run. We still downplay what's happening in faraway countries. And we discount too heavily the problems we'll leave for new generations. Without a broader perspective—without realizing that we're all on this crowded world together—governments won't properly prioritize projects that are long-term in a political perspective, even if a mere instant in the history of our planet. Knowing all we owe to past generations, it would be shameful if we weren't good ancestors and left a depleted heritage and damaged planet to our descendants.

The two-year-old girl who expressed wonder and joy at her first glimpse of the ocean is now forty-five years old. She has children of her own. Those young people are coming of age in a world that is at once wondrous and challenged, exhilarating and frightening, a world of potential and uncertainty, trembling, majestic, unpredictable and predictable, mysterious. A world to celebrate, and to preserve.

ACKNOWLEDGMENTS

We thank Lace Riggs, Dorota Grabowska, Marta Zlatić, John Mather, and Magdalena Lenda for their willingness to be interviewed.

NOTES

Chapter I: Disciplined Wonder

5 report of the Edelman Trust Barometer: 2024 Edelman Trust Barometer, https://www.edelman.com/trust/2024/trust-barometer.

5 study published in *Nature*: "Trust in Scientists and their Role in Society Across 68 Countries," *Nature,* published online January 20, 2025, https://doi.org/10.1038/s41562-024-02090-5.

8 "I made straight for Heligoland": Werner Heisenberg, *Physics and Beyond,* trans. Arnold J. Pomerans (New York: Harper and Row, 1971), pp. 60–61.

9 "I belong in the ranks of those": See Ernst P. Fischer, *Aristotele, Einstein e gli altri* (Milan: Raffaello Cortina Editore, 1977), pp. 270–71. Also see M. Bersanelli and M. Gargantini, *Galileo to Gell-Mann: The Wonder That Inspired the Greatest Scientists of All Time* (Philadelphia: Templeton Press, 2009), p. 9.

10 "Among these streaks": Letter to Henry Oldenburg, first secretary of the Royal Society, September 7, 1674, translated in C. Dobell, *Antony van Leeuwenhoek and His Little Animals* (New York: Russell and Russell, 1958), pp. 109–110, https://www.ncbi.nlm.nih.gov/pmc/articles/PMC4360124/#RSTB20140344C4.

10 "Any solid lighter": Archimedes, "On Floating Bodies" (ca. 250 BC), in *The Works of Archimedes,* ed. T. L. Heath (Cambridge: Cambridge University Press, 1897), book 1, prop 5.y.

13 "spandrels": An architectural term describing the decorative area between the curved top of an arch and the roof above it. The metaphor was introduced into biology by S. J. Gould and R. Lewontin to describe features of an organism that are by-products of evolution rather than of direct survival benefit.

14 "one feels tense": Irvin Yalom, *Existential Psychotherapy* (New York: Basic Books, 1980), pp. 462–63.

Chapter II: Why Science?

30 *Enûma Elish*: The *Enuma Elish* can be found in *The Babylonian Genesis,* trans. and ed. Alexander Heidel (Chicago: University of Chicago Press, 1942).

31 Plato argued that a Creator: The origin of the universe according to Plato can be found in Plato, *Timaeus,* trans. Benjamin Jowett, in *Great Books of the Western World* (Chicago: Encylopaedia Britannica, 1952), pp. 447–48.

35 "administers the Pleasures of Science": Sir Richard Steele, quoted in Judy Egerton, *Wright of Derby* (New York: Metropolitan Museum of Art, 1990), p. 54.

Chapter III: A Day in the Life

36 largest computer simulations of brains: Intel computer simulating brain: https://www.cnet.com/tech/computing/intel-packs-8-million-digital-neurons-onto-brain-like-pohoiki-beach-computer-loihi-chips.

37 "Living in severe poverty": Jennifer Michalowski, "Personal Pursuits," *Brain Scan* (Fall 2022). See also https://mcgovern.mit.edu/2022/09/09/personal-pursuits/.

38 "Despite everything around me": This and all other quotes by Riggs come from AL interviews with her on July 24, 2023, and August 18, 2023.

Chapter IV: How Scientists Think

59 "If a man will begin with certainties": Francis Bacon, *The Advancement of Learning* (1605), book 5, section 8 (London: Cassell & Company, 1893), https://www.gutenberg.org/files/5500/5500-h/5500-h.htm.

63 "no level of alcohol consumption is safe for our health": https://www.who.int/europe/news/item/04-01-2023-no-level-of-alcohol-consumption-is-safe-for-our-health#:~:text=The%20W.

64 "The human understanding resembles not a dry light": Francis Bacon, *Novum Organum* (1620), ed. Joseph Devey, book 1, section 49 (New York: P. F. Collier and Son, 1911), p. 26.

66 "As in reality such masses exist quietly": Lev Davidovich Landau, "On the Theory of Stars," *Physicalische Zeitschrift der Sowjetunion* 1 (1932): 285–88.

67 "The distinctive ability of a scientific discoverer": Michael Polanyi, *Personal Knowledge* (Chicago: University of Chicago Press, 1958), p. 302.

68 "Certainly a passion for what I'm working on": Thorne interviewed by AL, June 3, 2019.

69 "I have no great quickness": Charles Darwin, *The Autobiography of Charles Darwin,* ed. Nora Barlow (New York: Norton, 1969), p. 114.

72 "You go down there": Quotes from Grabowska are from AL interviews with them on October 18, 2021, and September 11, 2023.

74 "the whole of science is nothing more than": Albert Einstein, first published

in the *Journal of the Franklin Institute* 221, no. 3 (March 1936). See also Albert Einstein, "Physics and Reality," in *Ideas and Opinions* (New York: The Modern Library, 1994), p. 319.

82 "I was so interested in what I was doing": Evelyn Fox Keller, *A Feeling for the Organism* (New York: Freeman, 1983), p. 70. All of McClintock's direct quotes come from interviews with Keller, recorded in this book.

84 "My mother used to put a pillow": Keller, *A Feeling for the Organism,* p. 20.

84 "just thinking about things": Keller, *A Feeling for the Organism,* p. 22.

84 "My father tells me that at the age of five": Keller, *A Feeling for the Organism,* p. 22.

84 "I would solve some of the problems": Keller, *A Feeling for the Organism,* p. 26.

84 "Here was a dividing line": Keller, *A Feeling for the Organism,* p. 33.

84 "There was not that strong necessity": Keller, *A Feeling for the Organism,* p. 34.

85 "At this stage": Keller, *A Feeling for the Organism,* p. 72.

86 "Well, you know, when I look at a cell": Keller, *A Feeling for the Organism,* p. 69.

86 "the more I worked with [the chromosomes]": Keller, *A Feeling for the Organism,* p. 117.

86 "I was really quite petrified": Keller, *A Feeling for the Organism,* p. 115.

88 "I start with the seedling": Keller, *A Feeling for the Organism,* p. 198.

Chapter V: What Gets Them Started?

90 "my taste for natural history": Charles Darwin, *The Autobiography of Charles Darwin,* ed. Francis Darwin (London: John Murray, Albemarle Street, 1892), pp. 6, 9, 10.

91 "I was simply entranced by chemical phenomena": Irwin Abrams, *The Nobel Peace Prize and the Laureates* (Boston: G. K. Hager, 1988), p. 197.

92 "I was eight years old when": Joachim Frank, Nobel Prize autobiography, https://www.nobelprize.org/prizes/chemistry/2017/frank/facts/.

93 "though physically defeated": All quotes are from Shuji Nakamura, Nobel Prize biography, https://www.nobelprize.org/prizes/physics/2014/nakamura/biographical/.

93 "This early responsibility": Carol W. Greider, Nobel Prize autobiography, https://www.nobelprize.org/prizes/medicine/2009/greider/facts/.

95 "I can still remember a big board": Interview with AL on May 18, 1988, from Alan Lightman and Roberta Brawer, *Origins: The Lives and Worlds of Modern Cosmologists* (Cambridge, MA: Harvard University Press, 1990), pp. 400–401.

95 "It was love at first sight": Interview in *USC Today,* September 24, 2020, https://today.usc.edu/usc-researchers-childhood-start-career-in-science/.

96 "this experience led me to make a decision": Tu Youyou, Nobel Prize lecture, December 7, 2015, https://www.nobelprize.org/prizes/medicine/2015/tu/biographical/.

98 "Towards the end of high school": This and all of Marta Zlatić's quotes are from an AL interview with her on October 30, 2023.

106 "I saw what was meant by infinity": John C. Mather, 2006 Nobel Prize autobiography, https://www.nobelprize.org/prizes/physics/2006/mather/biographical/.

106 "In high school, I poked around": All Mather quotes unless otherwise noted are from an AL interview with him on September 28, 2023.

108 "I liked all three of them immediately": Mather, 2006 Nobel Prize autobiography.

109 "I was hoping to go into a new field of study": Mather, 2006 Nobel Prize autobiography.

110 "Tall, thin and bespectacled": John Boslough, "A Nobel Glow Lights Up Stockholm," *The Aspen Times,* December 26, 2006.

110 "Mather was infuriated": Carolyn Sayre, "Wild and Crazy Nobel Guys," *Time,* October 8, 2006.

112 "I extend special thanks": John Mather and John Boslough, *The Very First Light: The True Inside Story of the Scientific Journey Back to the Dawn of the Universe* (New York: Basic Books, 1996), acknowledgments.

Chapter VI: What Keeps Them Going?

114 "I don't see that it makes any point": Richard Feynman, *The Pleasure of Finding Things Out* (Cambridge, MA: Helix Books, 1999), p. 12.

115 "It's not hard to stay interested and curious": Carolyn Bertozzi, Nobel Prize interview, December 6, 2022, https://www.nobelprize.org/prizes/chemistry/2022/bertozzi/interview/.

115 "As the clock went past midnight": James Watson, *The Double Helix* (New York: New American Library, 1968), p. 118.

115 "The story of telomerase discovery": Carol Greider, Nobel Prize lecture, December 9, 2009.

116 "We have all the telescopes at our disposal": Interview with AL, December 1, 2023.

116 "Beginning in 2017": Email to AL, November 12, 2023.

117 "I am more sorry than I can tell you": Henrietta Leavitt to Edward Pickering, May 13, 1902, Harvard Archives, HCO Correspondence.

118 "I kept getting my charge": Interview with AL, November 21, 2021.

118 "I happened to walk into a basement": Freeman Dyson, "Science as a Craft Industry," *Science,* May 15, 1998.

119 "The joy of understanding": Roald Hoffmann and Jean-Paul Malrieu, "Simulation vs. Understanding: A Tension, in Quantum Chemistry and Beyond," *Angewandte Chemie* 59 (2020): 13694–710.

119 "As a working researcher": Email to AL, November 6, 2023.

120 "By the time the hour-and-a-half train journey": Watson, *The Double Helix*, p. 77 Kindle edition.

121 In his autobiography: Luis W. Alvarez, *Alvarez: Adventures of a Physicist* (New York: Basic Books, 1987).

122 "We were extremely lucky": Dennis Normale, "Shinya Yamanaka, Modest Researcher," *Science* 319 (February 1, 2008): 562.

123 "I like the act of measurement": Interview with AL, November 20, 2021.

124 "In science, and particularly in physics": Werner Heisenberg, *Physics and Beyond,* trans. Arnold J. Pomerans (New York: Harper and Row, 1971), p. 18.

127 "I had never been really at home": Heisenberg, *Physics and Beyond,* p. 17.

127 "Robert's references to Malebranche": Heisenberg, *Physics and Beyond,* pp. 11–12.

128 "looked like a simple farm boy": Max Born, *My Life* (New York: Scribner, 1978), p. 212. Heisenberg's 1973 lecture at Caltech was attended by AL, then a graduate student there.

131 "It must have driven him to utter despair": Victor Weisskopf, introduction to Elisabeth Heisenberg, *Inner Exile: Recollections of a Life with Werner Heisenberg,* trans. S. Cappellarii and C. Morris (Boston: Birkhauser, 1984), p. xiii.

131 "he would have abandoned his friends": Heisenberg, *Inner Exile,* p. 67.

131 "Toward the end of 1941": Heisenberg, *Physics and Beyond,* pp. 180–81.

133 "[Before the war], one could love one's work": Meitner to James Franck, March 1958, James Franck Papers, Joseph Regenstein Library, University of Chicago, quoted in Ruth Lewin Sime, *Lise Meitner: A Life in Physics* (Berkeley: University of California Press, 1996), p. 375.

135 "My father taught me how to identify plants": This quote and all others from Lenda, unless otherwise noted, are from AL interview with her on October 9, 2023.

136 "My idea of collaboration": Personal statement in ML's résumé.

141 "Never more shall I escape": Walt Whitman, "Out of the Cradle Endlessly Rocking" (1859), later incorporated in *The Leaves of Grass.*

Chapter VII: Patterns of Scientific Discovery

147 "As soon as I saw Balmer's formula": *Niels Bohr: A Centenary Volume,* ed. A. P. French and P. J. Kennedy (Cambridge, MA: Harvard University Press, 1985), p. 43.

147 "so striking that I dropped everything": Quoted in Evelyn Fox Keller, *A Feeling for the Organism* (New York: Freeman, 1983), p. 124.

148 The story of continental drift: The rejection of clues that violate current scientific theories until the emergence of a comprehensive new theory is discussed in the paper "When Do Anomalies Begin?" by Alan Lightman and Owen Gingerich, *Science* 255 (February 7, 1991): 690.

150 "In visualizing the cycle mechanism": Hans Krebs, *Reminiscences and Reflections* (Oxford: Clarendon Press, 1981), p. 118.

150 "came out just from playing with the equations": Interview with Thomas Kuhn, American Institute of Physics, 1963, https://www.aip.org/history-programs/niels-bohr-library/oral-histories/4575-3.

153 When nuclear physicist Lise Meitner was a child: Anecdote about Lise Meitner and her grandmother is from Ruth Lewin Sime, *Lise Meitner: A Life in Physics* (Berkeley: University of California Press, 1996), p. 5.

155 "I looked at the stars in the sky": All of Govind Swarup's direct quotes are from *Annual Reviews of Astronomy and Astrophysics* 59 (2021): 1–19, unless otherwise noted.

157 "the Father of Radio Astronomy in India": See, for example, Mehr Gill, "Explained: Who was Govind Swarup, the pioneer of radio astronomy in India?," *The Indian Express,* September 20, 2020, https://indianexpress.com/article/explained/govind-swarup-radio-astronomy-india-6588222/.

160 "always full of innovative ideas": Dhruba J. Saikia, "Govind Swarup," *Resonance,* July 2021.

161 "his ready and infectious smile": R. Ramachandran, "Govind Swarup (1929–2020): Star Among Astronomers," *Frontline,* September 26, 2020, https://frontline.thehindu.com/other/obituary/star-among-astronomers/article32632879.ece.

161 "created an environment": Goutam Chattopadhyay, *The Times of India,* September 9, 2020.

162 "At that time, I was very concerned": Ramachandran, "Govind Swarup (1929–2020)."

163 "I consider this a bigger achievement": Ramachandran, "Govind Swarup (1929–2020)."

Chapter VIII: The Ethics of Science and Responsibilities of Scientists

165 "his greatest scientific discovery": This and all quotes from Heisenberg and Weizsäcker come from Werner Heisenberg, *Physics and Beyond,* trans. Arnold J. Pomerans (New York: Harper and Row, 1971), pp. 194–200.

175 "to draw out of them": Francis Bacon, *The New Atlantis* (Oxford: The Clarendon Press, 1915), p. 45.

176 "freely and generously": Benjamin Franklin, *Autobiography of Benjamin Franklin* (1791), ch. 12, "Defense of the Province."

176 "for the better warming of rooms": Franklin, *Autobiography of Benjamin Franklin* (1791), ch. 12, "Defense of the Province."

183 Eric Schmidt, former CEO of Google: Eric Schmidt, "How We Can Control AI," *The Wall Street Journal,* January 26, 2024.

ILLUSTRATION CREDITS

Page

29 *Pleiades,* photograph by JoJan, CC BY 4.0,Wikimedia Commons

30 *Shaft of the Dead Man,* Glasshouse images/Alamy stock photo

34 *A Philosopher Giving that Lecture on the Orrery,* painting by Joseph Wright, public domain

45 *Neuron and axon,* Dhp1080, Wikimedia Commons

57 Mapping Specialists Ltd.

NOTES ABOUT THE AUTHORS

Alan Lightman is an American physicist, writer, and social entrepreneur with a PhD in physics from Caltech. He is the recipient of six honorary degrees. He has served on the faculties of Harvard and the Massachusetts Institute of Technology (MIT) and was the first person at MIT to receive dual faculty appointments in science and in the humanities. He is currently professor of the practice of the humanities at MIT. Lightman's work in astrophysics concerns black holes, stellar dynamics, and cosmic radiation processes. His essays about the intersection of science, culture, and philosophy have appeared in *The Atlantic, Harper's Magazine, The New Yorker,* and many other publications. He is the author of numerous books, both nonfiction and fiction, including *Einstein's Dreams,* an international bestseller, and *The Diagnosis,* a finalist for the National Book Award in fiction. Lightman is the host of the public television series *SEARCHING: Our Quest for Meaning in the Age of Science.* In August 2023, he was appointed a member of the United Nation's Scientific Advisory Board, to advise the secretary general about human issues associated with new developments in science and technology.

Martin Rees grew up in rural surroundings on the border of England and Wales and afterward studied at Cambridge University, specializing in astronomy and cosmology. After a peripatetic few years, he returned to Cambridge, where he later became director of the Institute of Astronomy and master of Trinity College. His main astronomical interests have been galaxy formation, cosmic jets, and black holes, along with more speculative topics: whether we live in a multiverse, and the prospects of detecting extraterrestrial life. He has received numerous international awards for his research. In addition to his research publications, he has written extensively for a general readership.

As president of the UK's Royal Society, a member of the House of Lords, and through other international bodies, he has been increasingly concerned about the impact of powerful new technologies on the world's expanding and more demanding population: How can we harness science's benefits while minimizing the downsides?

A NOTE ON THE TYPE

This book was set in Adobe Garamond. Designed for the Adobe Corporation by Robert Slimbach, the fonts are based on types first cut by Claude Garamond (ca. 1480–1561).

Composed by North Market Street Graphics,
Lancaster, Pennsylvania

Designed by Michael Collica